# ELEMENTARY TREATISE

ON

# DETERMINANTS

WITH THEIR APPLICATION TO

*SIMULTANEOUS LINEAR EQUATIONS*

*AND ALGEBRAICAL GEOMETRY.*

BY

CHARLES L. DODGSON, M.A.

STUDENT AND MATHEMATICAL LECTURER OF CHRIST CHURCH, OXFORD.

London:

MACMILLAN AND CO.

1867.

**British Library Cataloguing-in-Publication Data**
A catalogue record for this book is available from
the British Library

# Lewis Carroll

Lewis Carroll was born Charles Lutwidge Dodgson at the parsonage in Daresbury, Cheshire, England on 27th January 1832. His father was a brilliant mathematician, who neglected an academic career in favour of the clergy, eventually becoming the Archdeacon of Richmond.

One of eleven siblings, Dodgson was a precocious child, apparently reading John Bunyan's Christian allegory *The Pilgrim's Progress* by the age of seven. In 1843, the Dodgson family moved to North Yorkshire, and a year later the twelve-year old Lewis was sent to Richmond Grammar School. Two years later, he moved to Rugby School in Warwickshire. Despite an unhappy three years there, marked by illness and bullying, Dodgson revealed himself to be exceptionally gifted, particularly in mathematics.

Dodgson left Rugby towards the end of 1849 and matriculated at Oxford University in May of 1850 as a member of his father's old college, Christ Church. He went into residence eight months later, in the same week that his mother died from "inflammation of the brain."

At Oxford, despite a tendency towards procrastination, Dodgson excelled. He earned a first-class degree in mathematics in 1854, as well as a second-class degree in Classics, and completed his Masters three years later. Once more following in his father's footsteps, Dodgson was appointed as a lecturer of mathematics at Oxford 1856, a position he would go on to hold until 1881. Teaching provided Dodgson with a steady income, and over his career he would publish a number of successful mathematical textbooks, including *Two Books of Euclid* (1860), *Elementary Treatise on Determinants* (1867), *Examples in Arithmetic* (1874), and *Curiosa Mathematica, Part I: A New Theory of Parallels* (1888).

From a young age, Dodgson had penned poetry and short stories, and it was shortly after graduating that he began to write more seriously. Between 1854 and 1857, his work appeared in national publications such as *The Comic Times* and *The Train*, as well as smaller magazines like the *Whitby Gazette* and the *Oxford Critic*. In 1856, he published his first piece – a romantic poem called "Solitude" – under the pen name that he would eventually make him famous: Lewis Carroll.

Around this same time, Dodgson made the acquaintance of the new dean of Christ Church, Henry Liddell. During the late 1850s, Dodgson became close friends with Liddell's wife and young children, most notably their daughter, Alice. On an 1862 rowing expedition with the group, Dodgson came up with the outline of the story which would later become *Alice's Adventures in Wonderland*. Having recited the story to the young Alice Liddell, who had begged him to write it down, in 1864 Dodgson presented her with a handwritten, illustrated manuscript entitled *Alice's Adventures Under Ground*.

The young Alice's enthusiasm encouraged Dodgson to pursue publication, and he quickly received interest from the London-based house Macmillan. After the possible alternative titles *Alice Among the Fairies* and *Alice's Golden Hour* were rejected, the work was finally published as *Alice's Adventures in Wonderland* in 1865 under the Lewis Carroll pen-name, with illustrations by John Tenniel.

Telling the tale of a girl named Alice who falls down a rabbit hole into a fantasy world populated by surreal and anthropomorphic creatures, the book was a huge commercial success, and Carroll was swamped with fan mail. Despite initial lukewarm reviews, it would become increasingly popular over the following years, and is today regarded as one of the finest examples of both the literary nonsense genre and children's literature more generally.

Despite the success and financial reward brought by *Alice's Adventures in Wonderland*, Carroll maintained his position at Christ Church. Having become a deacon at the cathedral some years earlier – but for reasons largely unknown, neglected to continue on into the priesthood - he continued to lead services, and to teach mathematics at the university. In 1867, Carroll travelled through Europe, returning shortly before the death of his father, which plunged him into a deep depression. In 1869, Carroll published his first major collection as *Phantasmagoria*, following it seven years later with the epic nonsense poem regarded as his last great work, "The Hunting of the Snark". In 1871, Carroll's sequel to *Alice's Adventures in Wonderland, Through the Looking Glass and What Alice Found There*, appeared. Darker in tone, it included arguably his most famous poem, "Jabberwocky".

In 1881, Carroll resigned his lectureship at Oxford in order to focus on his writing. Over the next decade, he continued to produce fiction, as well as works on philosophy, mathematics and political theory. He died on 14th January 1898, two weeks from turning sixty-six, and now lies buried with many of his siblings at The Mount cemetery in Guildford, Surrey, England.

Since his death, Carroll's legacy has been a complicated one. A deeply private man, prone to nerves when questioned publicly, Carroll destroyed many of his personal papers, and over time various theories have abounded relating to his inner character. Some have suggested that he suffered from mental illness, or was a drug abuser, and newspapers such as the *Daily Mail* have suggested he possessed an unhealthy interest in children. However, in recent times there has been a concerted effort by scholars to do away with the 'Carroll Myth', and to present a more measured and less sensationalist account of the man. There has also emerged an increasing focus on his lesser-known works, and a focus on assessing him as not just an author of children's literature, but also as a gifted mathematician, philosopher, photographer, and even inventor.

# PREFACE.

OF the seventy Propositions contained in the following treatise, ten are substantially taken from Baltzer's treatise on Determinants; also the Geometrical Tests, given in Chapter VIII, are to be found in most works on Algebraical Geometry: the rest of the matter is, so far as I know, original, and consists of a series of Propositions which the object I had in view obliged me to introduce. That object was to present the subject as a continuous chain of argument, separated from all accessories of explanation or illustration, a form which I venture to think better suited for a treatise on exact science than the semi-colloquial semi-logical form often adopted by Mathematical writers. I say 'semi-logical' advisedly, for nothing is more easy than to forget, in an argument thus interwoven with illustrative matter, what has, and what has not, been proved.

With this object in view I have introduced all such explanation and illustration as seemed necessary for a beginner, either in the form of foot-notes, or, where that would have occupied too much room, of Appendices.

New words and symbols are always a most unwelcome addition to a Science, especially to one already burdened with an enormous vocabulary, yet I think the Definitions given of them will be found to justify their introduction, as the only way of avoiding tedious periphrasis. The symbols employed to represent the single elements of a Determinant, $\left(1\backslash 2, 1\backslash 3, \&c.\right)$

require perhaps a word of apology, and it may be well to enumerate those already in use, and to point out what seem to be their chief defects.

We may commence with $\left\{ \begin{matrix} a_1, & b_1, & \ldots\ldots \\ a_2, & b_2, & \ldots\ldots \end{matrix} \right\}$, where the change of *letter* indicates a change of *column*, and the change of *subscript* a change of *row*. Now the properties of Determinants, relating to columns, being always convertible into properties relating to rows, and vice versâ, it was a sufficient objection to this system of notation, that it represented things distinctly analogous by methods so different, and it was properly superseded by the notation introduced by Leibnitz, $\left\{ \begin{matrix} a_{1,1}, & a_{1,2}, & \ldots\ldots \\ a_{2,1}, & a_{2,2}, & \ldots\ldots \end{matrix} \right\}$, where the changes, both of column and row, are alike denoted by subscripts. But it seems a fatal objection to this system that most of the space is occupied by a number of $a$'s, which are wholly superfluous, while the only important part of the notation is reduced to minute subscripts, alike difficult to the writer and the reader. It was almost an obvious improvement on this system to raise the subscripts into the line, and omit the $a$'s altogether, as suggested by Baltzer, thus — $\left\{ \begin{matrix} (1,1), & (1,2), & \ldots\ldots \\ (2,1), & (2,2), & \ldots\ldots \end{matrix} \right\}$, and this system, though tedious for writing, might serve very well, were it not for its liability to be confused with the notation, common in Plane Algebraical Geometry, by which (1,1) denotes the Point $x=1$, $y=1$. The symbol $1\backslash 1$, which I have ventured to suggest as an emendation on this last, will be found, I have great hopes, sufficiently simple, distinct, and easy to be written. I have turned the symbol towards the left, in order to avoid all chance of confusion with $\int$, the symbol for integration.

I proceed to make a few introductory remarks on the various portions of the book, taken in order.

*Chap. II. Def. I.* I am aware that the word 'Matrix' is already in use to express the very meaning for which I use the word 'Block'; but surely the former word means rather the mould, or form, into which algebraical quantities may be introduced, than an actual assemblage of such quantities; for instance, $\frac{(\quad)\times(\quad)}{(\quad)}$ would deserve the name, rather than $\frac{(a+b)\times(c+d)}{(e+f)}$.

*Chap. II. Def. I, VIII.* Those who have read the chapters on Determinants in Mr. Todhunter's 'Theory of Equations' will notice that the meanings of the words 'Element' and 'Constituent' are here transposed: as to the former, I have only returned to Baltzer's nomenclature; and the word 'Constituent' seems to me more expressive than his word 'Term'.

*Chap. III.* A complete analysis of a system of simultaneous Linear Equations has always appeared to me to be a desideratum in Algebra: the subject is only touched on in Baltzer; a more complete attempt will be found in Peacock's Algebra, but I have nowhere seen anything like an exhaustive analysis. This chapter aims at furnishing this, but it has been so often altered and re-written that I put it forth at last, hoping, rather than expecting, that it will be found complete and satisfactory.

*Chap. VII.* This chapter will also, I hope, fulfil my aim at furnishing an *exhaustive* analysis of such properties of the Loci here considered, as can be conveniently exhibited in the form of Determinants. I had added propositions concerning the Line in Solid Geometry, but these I omit, believing that its properties are more simply investigated by other methods.

*Appendix II. Section* 4. This process, though extremely convenient where no ciphers, or where one or two at most, occur in the interior of a Block, nevertheless fails entirely, it must be admitted, where they occur in larger numbers: I therefore offer it merely as a fanciful addition to the processes already in use, which may in some cases lessen the labour of computation.

*Appendix V.* I am doubtful whether this process will ever prove of much practical use: still I think cases might arise, where in the course of a problem an algebraical function is proved to vanish, and where, by throwing it into the form of a Determinant, and so forming a set of simultaneous Equations, whose consistency depends on its vanishing, new and curious properties of the function under consideration might be evolved.

The formulæ given at the end of the book are so arranged that the student may, by covering one or more of the columns on the right hand, test for himself his knowledge of the theorems from which they are taken.

CH. CH. OXFORD,
Oct. 31, 1867.

# CORRIGENDA.

P. 36. l. 10. *for*

and since, by hypothesis, $V \neq 0$, these Equations may be divided throughout by $V$, and written

$$x_1 = \frac{D_1}{V}, \quad -x_2 = \frac{D_2}{V}.$$

*read*

and, dividing these Equations throughout by $V$,

$$x_1 = \frac{D_1}{V}, \quad -x_2 = \frac{D_2}{V};$$

and since, by hypothesis, $V \neq 0$, these values are both finite.

P. 50. l. 15. *for* $B = 0$ *read* $V = 0$.

P. 51. l. 3. *for* $\|B\| = 0$ *read* $\| V \| = 0$.

# CONTENTS.

# CHAPTER I.

## *LAWS OF ARRANGEMENT.*

### Definitions.

I.

A set of different numerals, arranged in an ascending order, is said to be **orderly arranged**: but if there be among them 2, of which the second is less than the first, the set is said to contain a **derangement** *.

II.

If 2 numbers be both even, or both odd, they are said to be **similar**; if otherwise, **dissimilar**.

### Proposition I. Th.

If there be a set of different numerals, arranged in any order, and if one of them be made to pass over the next $r$ of them, either way: the number of derangements is increased, or diminished, by a number similar to $r$.

If it be made to pass over *one*, the number is increased, or diminished, by unity;

∴ if over *two*, by an even number;

---

* *Def. I.* Thus the set 12346789 is *orderly* arranged; but the same set placed thus, 43186972, contains one derangement on account of the 4 and 3, another on account of the 4 and 1, another on account of the 8 and 7, and so on.

∴ if over *three*, by an odd number; and so on.

Therefore, if there be a set, &c. Q. E. D*.

## PROPOSITION II. TH.

If there be a set of different numerals, and if 2 of them be interchanged: the number of derangements is increased, or diminished, by an *odd* number.

Call the 2 numerals, $\alpha$, $\beta$; and let there be $r$ numerals between them;

firstly, let $\alpha$ be made to pass over these $r$ numerals;

then the number of derangements is increased, or diminished, by a number similar to $r$; (PROP. I.

secondly, let $\beta$ be made to pass over $\alpha$ and over these $r$ numerals;

then the number of derangements is thereby increased, or diminished, by a number similar to $\overline{r+1}$; (PROP. I.

∴ it is ultimately increased, or diminished, by the sum or difference of 2 dissimilar numbers;

i. e. it is increased, or diminished, by an *odd* number.

Therefore, if there be a set, &c. Q. E. D †.

## PROPOSITION III. TH.

If there be a set of pairs of numerals, in which the antecedents are all different, as also are the consequents; and if they be arranged, firstly in order of antecedents, and secondly in order of consequents: the number of derangements among the consequents in the first case, and the number of derangements among the antecedents in the second case, are equal.

---

* *Prop. I.* In the set 43186972, let us make the 7 pass over the preceding 3 numerals. By passing it over the 9, a derangement is lost, i. e. the number of derangements is diminished by unity; by passing it over the 6, a derangement is gained, i. e. the number is what it was at first, i. e. it is increased by zero, which is *even*; by passing it over the 8, a derangement is lost, i. e. the number is diminished by unity, which is *odd*; and so on.

† *Prop. II.* In the set 43186972, let us interchange the 1 and the 7. By passing the 1 over the intermediate 3 numerals, 3 derangements are gained, i. e. the number of derangements is increased by a number similar to 3; and the set now stands thus, 43869172; by passing the 7 over the 1 and over the same 3 numerals, two derangements are gained and two lost, i. e. the number is increased by zero, which is similar to (3+1); hence it is on the whole increased by a number similar to the *sum* of 3 and (3+1), and, as these are *dissimilar* numbers, their sum is *odd*.

Let the pairs be so placed that the antecedents are orderly arranged, and let 2 of them be selected, and call them $(H, r)$, $(K, s)$;

∴ $H < K$;

now, if these 2 pairs contain a derangement of consequents, $r > s$;

∴ when the pairs are re-arranged in order of consequents, these 2 will stand in the order $\ldots(K, s)\ldots(H, r)\ldots$;

∴ they will then contain a derangement of antecedents;

but if these 2 pairs do not contain a derangement of consequents, $r < s$;

∴ when the pairs are re-arranged in order of consequents, these 2 will stand in the order $\ldots(H, r)\ldots(K, s)\ldots$;

∴ they will then not contain a derangement of antecedents.

And the same thing may be proved for every other 2 pairs.

Therefore, if there be, &c. Q. E. D *.

---

## Definition III.

If there be a set of pairs of numerals, in which the antecedents are all different, as also are the consequents; and if, when they are arranged in order of antecedents, the number of derangements among the consequents be *even*, or (which is the same thing) if, when they are arranged in order of consequents, the number of derangements among the antecedents be *even*: the set is said to be **of the even class**; if otherwise, **of the uneven class** †.

---

* *Prop. III.* Let the set be (1, 3), (3, 8), (4, 6), (7, 5), (8, 2), which is arranged in order of antecedents: if this be rearranged in order of consequents, it will stand thus:—(8, 2), (1, 3), (7, 5), (4, 6), (3, 8). Now let us select 2 of these pairs, (1, 3) and (7, 5); these, as they stand in the first arrangement, contain no derangement of consequents; hence, in the second arrangement, they preserve the same relative order, and so contain no derangement of antecedents. Again, let us select (3, 8) and (7, 5); these, as they stand in the first arrangement, contain a derangement of consequents; hence, in the second arrangement, they take the order (7, 5),....(3, 8), and so contain a derangement of antecedents. And so for every other 2.

† *Def. III.* Taking the set (1, 3), (3, 8), (4, 6), (7, 5), (8, 2), let us ascertain, by counting the derangements among the consequents, to which class it belongs. This may be conveniently done by observing, for each consequent in turn, how many of its predecessors are *greater* than it, since every instance of this will constitute a derangement: thus the 3 gives none, the 8 gives none, the 6 gives one, the 5 gives two, and the 2 gives four; hence there are *seven* derangements among the consequents. Again, let us arrange the set in order of consequents, (8, 2), (1, 3), (7, 5), (4, 6), (3, 8), and count

## PROPOSITION IV. TH.

If there be a set of pairs of numerals, in which the antecedents are all different, as also are the consequents; and if 2 of the antecedents, or 2 of the consequents, be interchanged: the class, to which the set belongs, is changed.

Let the set be arranged in order of antecedents, and let 2 of the consequents be interchanged;

then the number of derangements among them is increased, or diminished, by an *odd* number; (PROP. II.

i. e. if even, it becomes odd; if odd, even;

∴ the class, to which the set belongs, is changed;

hence, if the set be arranged in any order, and 2 of the consequents be interchanged, the class, to which the set belongs, is changed.

Similarly, if 2 of the antecedents be interchanged.

Therefore, if there be, &c. Q. E. D*.

## PROPOSITION V. TH.

If there be a set of $n$ pairs of numerals, in which the antecedents are a certain permutation of the numbers from 1 to $n$, as also are the consequents; and if one pair be erased: the class, to which the remaining set belongs, is the same as that of the original set, or different, according as the numerals in the erased pair are similar or dissimilar.

Let the set be arranged in order of antecedents; and call the pair that is to be erased $(H, k)$;

---

the derangements among the antecedents: thus the 8 gives none, the 1 gives one, the 7 gives one, the 4 gives two, and the 3 gives three; hence there are now *seven* derangements among the antecedents. Thus the set of pairs of numerals, tried by either test, is of the *uneven* class.

It should be observed that the class, to which a set of pairs of numerals belongs, is unaffected by the order in which they happen to be given.

* *Prop. IV.* In the set (arranged, for convenience, in order of antecedents), (1, 3), (3, 8), (4, 6), (7, 5), (8, 2), let us interchange the two consequents, 8 and 5; the set will thus become (1, 3), (3, 5), (4, 6), (7, 8), (8, 2). Now, by this interchange, the number of derangements among the consequents is diminished by three; i. e. from being *odd*, it becomes *even*; and the set of numerals is therefore transferred from the *uneven* to the *even* class.

firstly, let it be brought to the first place, by making it pass over the preceding $\overline{H-1}$ pairs;

then the number of derangements among the consequents is increased, or diminished, by a number similar to $\overline{H-1}$; (PROP. I.

and, since the consequent $k$ now precedes the $\overline{k-1}$ consequents which are less than it, there are now, by reason of this pair, $\overline{k-1}$ derangements among the consequents;

secondly, let the pair $(H, k)$ be erased;

then the number of derangements among the consequents is thereby diminished by a number similar to $\overline{k-1}$;

∴ it is ultimately increased, or diminished, by a number similar to the sum or difference of $\overline{H-1}$ and $\overline{k-1}$;

i. e. it is ultimately increased, or diminished, by an even, or odd, number, according as $\overline{H-1}$ and $\overline{k-1}$ are similar or dissimilar;

i. e. according as $H$ and $k$ are similar or dissimilar.

Therefore, if there be a set, &c. Q. E. D*.

---

* *Prop. V.* Let us take the following set, (arranged, for convenience, in order of antecedents), (1, 2), (2, 4), (3, 1), (4, 5), (5, 6), (6, 3); and let us select (4, 5) as the pair to be erased, in which the numerals are dissimilar. Firstly, let us bring it to the first place, so that the set now stands thus:— (4, 5), (1, 2), &c.; in doing this, we have made the consequent 5 pass over the preceding (4−1) consequents, and have thus increased, or diminished, the number of derangements among the consequents, by a number similar to (4−1). And since this consequent 5 now precedes all the lesser consequents, 1, 2, 3, 4, there are now, by reason of it, (5−1) derangements among the consequents. Next, let the pair (4, 5) be erased; then these (5−1) derangements are done away with, and the number of derangements is, on the whole, increased, or diminished, by a number similar to the sum, or difference, of (4−1) and (5−1), i. e. to the sum, or difference, of 4 and 5, and since they are *dissimilar* numbers, their sum, or difference, is *odd*; hence the class, to which the set belongs, is *changed.*

In this instance it will be found that the given set contains 5 derangements of consequents, and so is of the *uneven* class; and that the new set, (1, 2), (2, 4), (3, 1), (5, 6), (6, 3), contains 4, and so is of the *even* class.

# CHAPTER II.

## *ANALYSIS OF DETERMINANTS.*

### DEFINITIONS.

I.

If $mn$ quantities be so placed as to form $m$ rows and $n$ columns: they are said to form a **Block**; and the $mn$ quantities are called the **Elements** of such a Block.

II.

A square Block of $n^2$ Elements is said to be **of the n$^{th}$ degree.**

III.

An oblong Block containing $m$ rows and $n$ columns, or $m$ columns and $n$ rows, where $m$ is greater than $n$, is said to be **of the length m**, and **of the breadth n.**

IV.

In an oblong Block, the rows, if they be longer than the columns, or the columns, if they be longer than the rows, are called the **longitudinals** of the Block: and the others, its **laterals.**

## V.

If, in a given Block, any rows, and as many columns, be selected: the square Block formed of their common Elements is called a **Minor** of the given Block *.

Hence any single Element of a Block, being common to one row and one column, is a Minor of it.

## VI.

If $n$ be that dimension of a Block which is not greater than the other: its Minors of the $n^{\text{th}}$ degree are called its **principal Minors**; those of the $\overline{n-1}|^{\text{th}}$ degree its **secondary Minors**, and so on †.

Hence a square Block is its own principal Minor.

## VII.

If, in a square Block, any rows, and as many columns, be selected: the Minor formed of their common Elements, and the Minor formed of the Elements common to the other rows and columns, are said to be **complemental** to each other ‡.

---

## CONVENTIONS.

## I.

Let it be agreed to represent the Elements of a square Block by symbols of the form $h\backslash k$, in which the first numeral indicates

---

* *Def. V.* Thus, in the Block $\left\{\begin{matrix} d & b & m & s \\ f & c & g & d \\ e & h & r & l \end{matrix}\right\}$, if we select the 2nd and 3rd rows, and the 2nd and 4th columns, we obtain the Minor $\left\{\begin{matrix} c & d \\ h & l \end{matrix}\right\}$.

† *Def. VI.* Thus, in the same Block, the Minors $\left\{\begin{matrix} d & b & m \\ f & c & g \\ e & h & r \end{matrix}\right\}$, $\left\{\begin{matrix} d & m & s \\ f & g & d \\ e & r & l \end{matrix}\right\}$, &c., are *principal* Minors; while $\left\{\begin{matrix} d & m \\ f & g \end{matrix}\right\}$, $\left\{\begin{matrix} d & s \\ f & d \end{matrix}\right\}$, $\left\{\begin{matrix} b & s \\ h & l \end{matrix}\right\}$, &c., are *secondary* Minors.

‡ *Def. VII.* Thus, in the Block $\left\{\begin{matrix} b & g & h & r \\ c & l & t & v \\ d & m & f & e \\ a & s & x & q \end{matrix}\right\}$, the Minors $\left\{\begin{matrix} b & g \\ c & l \end{matrix}\right\}$ and $\left\{\begin{matrix} f & e \\ x & q \end{matrix}\right\}$ are complemental to each other; as also are the Minors $\left\{\begin{matrix} c & v \\ a & q \end{matrix}\right\}$ and $\left\{\begin{matrix} g & h \\ m & f \end{matrix}\right\}$. Thus, again, the single Element $f$ and the Minor $\left\{\begin{matrix} b & g & r \\ c & l & v \\ a & s & q \end{matrix}\right\}$ are complemental to each other.

the row, and the second the column, to which the Element belongs. Thus, a Block of $m$ rows and $n$ columns may be represented thus :—

$$\left\{\begin{array}{lll} 1\backslash 1, & 1\backslash 2 & \cdots\cdots\ 1\backslash n \\ 2\backslash 1, & 2\backslash 2 & \cdots\cdots\ 2\backslash n \\ \vdots & \vdots & \qquad\ \ \vdots \\ m\backslash 1, & m\backslash 2 & \cdots\cdots\ m\backslash n \end{array}\right\}.$$

II.

And if it be required to represent 2 or more such Blocks, let them be distinguished by suffixing a certain letter to the symbol of each Block : e. g. :—

$$\left\{\begin{array}{ll} 1\backslash 1 \cdots\cdots & 1\backslash n \\ \vdots & \ \ \vdots \\ m\backslash 1 \cdots\cdots & m\backslash n \end{array}\right\}_a , \quad \left\{\begin{array}{ll} 1\backslash 1 \cdots\cdots & 1\backslash n \\ \vdots & \ \ \vdots \\ m\backslash 1 \cdots\cdots & m\backslash n \end{array}\right\}_b .$$

III.

And if it be required to represent an Element of such a Block by itself, let it be distinguished by the same suffix : e. g., $h\backslash k_a$ represents an Element of the Block

$$\left\{\begin{array}{ll} 1\backslash 1 \cdots\cdots & 1\backslash n \\ \vdots & \ \ \vdots \\ m\backslash 1 \cdots\cdots & m\backslash n \end{array}\right\}_a ;$$

$h\backslash k_b$ represents the corresponding Element of the Block

$$\left\{\begin{array}{ll} 1\backslash 1 \cdots\cdots & 1\backslash n \\ \vdots & \ \ \vdots \\ m\backslash 1 \cdots\cdots & m\backslash n \end{array}\right\}_b .$$

## DEFINITIONS (*continued*).

### VIII.

If there be a square Block of the $n^{\text{th}}$ degree, and if all possible products be made of its Elements, taken $n$ together, so that no product contain 2 Elements of the same row or of the same column; and if, representing the Elements of the Block by the symbols

$$\begin{Bmatrix} 1\diagdown 1 \ldots\ldots 1\diagdown n \\ \vdots \qquad\qquad \vdots \\ n\diagdown 1 \cdots\cdots n\diagdown n \end{Bmatrix},$$

each such product be affected with + or −, according as the set of pairs of numerals, corresponding to that product, be of the even or the uneven class: the sum of these products, thus affected, is called the **Determinant** of the Block. And each of these products is called a **Constituent** of the Determinant.

### IX.

The Constituent represented by the product $1\diagdown 1 . 2\diagdown 2 \ldots\ldots n\diagdown n$ is called the **Diagonal** of the Determinant *.

### X.

If a square Block be such that its Determinant vanishes, or if an oblong Block be such that the Determinant of every one of its principal Minors vanishes: in either case the Block is said to be **evanescent.**

---

* *Def. VIII, IX.* Thus, in the Block,

$$\begin{Bmatrix} b & g & h & r \\ c & l & t & v \\ d & m & f & e \\ a & s & x & q \end{Bmatrix},$$

the Diagonal is *blfq*, and the other Constituents are *blxe*, *bmtq*, *dstr*, &c. And to determine the *sign* of each Constituent, let us take as an example *dstr*; now this corresponds to the set of general symbols $3\diagdown 1 . 4\diagdown 2 . 2\diagdown 3 . 1\diagdown 4$, and since this is arranged in order of consequents and there are 5 derangements among the antecedents, it is of the *uneven* class, and so must be affected with the sign −.

## XI.

If there be a square Block, and if one of its Elements be selected; and if all the Constituents of its Determinant, which contain that Element, be collected together and formed into 2 factors, whereof that Element is one: the other factor is called the **determinantal coefficient** of that Element.

## XII.

If 2 square Blocks be such that each Element of the second is equal to the determinantal coefficient of the corresponding Element of the first: the second Block is said to be **adjugate** to the first.

---

CONVENTIONS (*continued*).

## IV.

Let it be agreed that the Determinant of a Block shall be represented by placing a perpendicular line on each side of it. Thus the Determinant of the Block

$$\begin{Bmatrix} a & b & c \\ d & e & f \\ g & h & k \end{Bmatrix}$$

will be represented by the symbol

$$\begin{vmatrix} a & b & c \\ d & e & f \\ g & h & k \end{vmatrix}$$

## V.

If a Block be represented by the symbol

$$\left\{\begin{matrix} 1\backslash 1 & \cdots\cdots & 1\backslash n \\ \vdots & & \vdots \\ n\backslash 1 & \cdots\cdots & n\backslash n \end{matrix}\right\}_a,$$

so that any Element of it is represented by a symbol of the form $h\backslash k_a$ : let it be agreed that the determinantal coefficient of that Element shall be represented by a symbol of the form $h\backslash k_A$.

## VI.

If there be 2 equally numerous sets of terms; and if each term of the one set be multiplied by the corresponding term of the other, and the products added: let it be agreed that this operation shall be denoted by placing the sign $\S$ between the symbols denoting the 2 sets.

Thus $(a_1, a_2, \ldots a_n) \S (b_1, b_2, \ldots b_n) = a_1 b_1 + a_2 b_2 + \ldots + a_n b_n$.

---

## Axioms.

### I.

Each Constituent of the Determinant

$$\begin{vmatrix} 1\backslash 1 & \cdots\cdots & 1\backslash n \\ \vdots & & \vdots \\ n\backslash 1 & \cdots\cdots & n\backslash n \end{vmatrix}$$

contains $n$ pairs of numerals, such that the antecedents are a certain permutation of the numbers 1 to $n$, as also are the consequents.

II.

If any row, or column, of a square Block be selected: each Constituent of the Determinant contains one term of that row or column.

Thus the Determinant $\left| \begin{matrix} 1\backslash 1 \cdots\cdots 1\backslash n \\ \vdots \qquad\quad \vdots \\ n\backslash 1 \cdots\cdots n\backslash n \end{matrix} \right|_a$

$$= 1\backslash 1_a \,.\, 1\backslash 1_A + 1\backslash 2_a \,.\, 1\backslash 2_A + \ldots\ldots + .\, 1\backslash n_a \,.\, 1\backslash n_A\,;$$

$$\text{or } = 1\backslash 1_a \,.\, 1\backslash 1_A + 2\backslash 1_a \,.\, 2\backslash 1_A + \ldots\ldots + .\, n\backslash 1_a \,.\, n\backslash 1_A\,;$$

$$\text{or } = h\backslash 1_a \,.\, h\backslash 1_A + h\backslash 2_a \,.\, h\backslash 2_A + \ldots\ldots + .\, h\backslash n_a \,.\, h\backslash n_A\,;$$

$$\text{or } = 1\backslash k_a \,.\, 1\backslash k_A + 2\backslash k_a \,.\, 2\backslash k_A + \ldots\ldots + .\, n\backslash k_a \,.\, n\backslash k_A\,.$$

Hence if, in a square Block, the Elements of any one row, or column, be multiplied by $v$; the Determinant of the new Block is equal to that of the first multiplied by $v$.

And if the Elements of any $q$ rows, or $q$ columns, or $r$ rows and $s$ columns, where $\overline{r+s} = q$, be multiplied by $v$; the Determinant of the new Block is equal to that of the first multiplied by $v^q$.

And if each Element of any row, or column, be the sum of $m$ terms: the Determinant may be expressed as the sum of $m$ Determinants. For one Determinant may be formed, such that its corresponding row, or column, consists of the first terms of these sums; another, such that its corresponding row, or column, consists of the second terms of these sums; and so on.

---

PROPOSITION I. TH.

The determinantal coefficient of any Element of a square Block is the Determinant of its complemental Minor, affected with + or − according as the numerals which constitute its symbol are similar or dissimilar.

Let the Block be represented by the symbol:—

$$\left\{\begin{matrix} 1\backslash 1 \cdots\cdots 1\backslash n \\ \vdots \qquad\quad \vdots \\ n\backslash 1 \cdots\cdots n\backslash n \end{matrix}\right\};$$

and call the selected Element $h\backslash k$;

then it is evident that the determinantal coefficient of $h\backslash k$ is the aggregate of all possible products of the Elements of its complemental Minor, taken $\overline{n-1}$ together, so that no product contain 2 Elements of the same row, or of the same column;

i. e. it is the aggregate of the Constituents of the Determinant of its complemental Minor; (DEF. VIII.

also the *sign* of each such product is + or −, according as the corresponding set of pairs of numerals, taken along with the symbol $h\backslash k$ itself, is of the even or of the uneven class; (DEF. VIII.

but if the symbol $h\backslash k$ be erased, the class, to which the remaining set belongs, is the same as that of the original set, or different, according as $h$ and $k$ are similar, or dissimilar; (CHAP. I. PROP. V.

∴ the *sign* of each such product follows the Determinant law, or reverses it, according as $h$ and $k$ are similar, or dissimilar;

∴ the determinantal coefficient of any Element $h\backslash k$ is the Determinant of its complemental Minor, affected with + or −, according as $h$ and $k$ are similar or dissimilar.

Q. E. D*.

## COROLLARIES TO PROP. I.

1.

If, in a square Block, any row, or column, be selected: the Determinant

---

* *Prop. I.* Thus the Determinant of the Block $\left\{\begin{matrix} a & b \\ c & d \end{matrix}\right\}$ is $(ad-bc)$; and that of the Block $\left\{\begin{matrix} a & b & c \\ d & e & f \\ g & h & k \end{matrix}\right\}$ is $(aek-ahf-bdk+bgf+cdh-cge)$. Here the Determinantal coefficient of $e$ is $(ak-cg)$, i. e. $\left|\begin{matrix} a & c \\ g & k \end{matrix}\right|$; and as $e$ corresponds to the symbol $2\backslash 2$, the numerals of which are *similar*, the sign of this Determinant ought to be +, and so we find it. Again, the Determinantal coefficient of $f$ is $(-ah+bg)$, i. e. $-\left|\begin{matrix} a & b \\ g & h \end{matrix}\right|$; and as $f$ corresponds to the symbol $2\backslash 3$, the numerals of which are *dissimilar*, the sign of this Determinant ought to be −, and so we find it.

of the Block may be resolved into terms, each consisting of one of the Elements of that row, or column, multiplied by the Determinant of its complemental Minor *.

2.

If, in a square Block, the Elements in any one row, or column, all vanish but one: the Determinant of the Block is the product produced by multiplying the Determinant of the complemental Minor of that Element by that Element itself, affected with + or −, according as the numerals in its symbol are similar or dissimilar.

3.

Hence, if that Element be unity, the Determinant of the Block is the Determinant of the complemental Minor of that Element, affected with + or −, as before.

## PROPOSITION II. TH.

If, in a square Block, 2 rows, or 2 columns, be interchanged: the Determinant of the new Block has the same absolute value as that of the first, but the opposite sign.

Let the Block be represented by

$$\left\{\begin{matrix} 1\backslash 1 \cdots\cdots 1\backslash n \\ \vdots \qquad\quad \vdots \\ n\backslash 1 \cdots\cdots n\backslash n \end{matrix}\right\}_a,$$

and first let 2 rows be interchanged; call them the $h^{\text{th}}$ and $k^{\text{th}}$ rows; and let the new Block be represented by

$$\left\{\begin{matrix} 1\backslash 1 \cdots\cdots 1\backslash n \\ \vdots \qquad\quad \vdots \\ n\backslash 1 \cdots\cdots n\backslash n \end{matrix}\right\}_b;$$

---

* *Prop. I. Cor.* 1. This gives us a simple method for computing the value of a Determinant arithmetically. Thus,

$$\begin{vmatrix} 3 & 1 & 2 & 4 \\ 4 & 5 & 2 & 3 \\ 3 & 1 & 3 & 2 \\ 4 & 2 & 1 & 3 \end{vmatrix} = 3\begin{vmatrix} 5 & 2 & 3 \\ 1 & 3 & 2 \\ 2 & 1 & 3 \end{vmatrix} - 1\begin{vmatrix} 4 & 2 & 3 \\ 3 & 3 & 2 \\ 4 & 1 & 3 \end{vmatrix} + 2\begin{vmatrix} 4 & 5 & 3 \\ 3 & 1 & 2 \\ 4 & 2 & 3 \end{vmatrix} - 4\begin{vmatrix} 4 & 5 & 2 \\ 3 & 1 & 3 \\ 4 & 2 & 1 \end{vmatrix}$$

$$= 3\left\{5\begin{vmatrix} 3 & 2 \\ 1 & 3 \end{vmatrix} - 2\begin{vmatrix} 1 & 2 \\ 2 & 3 \end{vmatrix} + 3\begin{vmatrix} 1 & 3 \\ 2 & 1 \end{vmatrix}\right\} - \&c. = 3\left\{35 + 2 - 15\right\} - \&c. = 3 \times 22 - \&c. = 66 - \&c.$$

$\therefore\quad h\backslash r_a = k\backslash r_b$, and $k\backslash r_a = h\backslash r_b$, $r$ taking any value $1 \ldots\ldots n$;

next let any Constituent of the Determinant of the original Block be selected; call it $1\backslash a_a \ldots\ldots h\backslash s_a \ldots\ldots k\backslash t_a \ldots\ldots$; and let it $= M$;

now $1\backslash a_a = 1\backslash a_b$ and so for the other factors of it, with the exception of $h\backslash s_a$ and $k\backslash t_a$ which respectively $= k\backslash s_b$ and $h\backslash t_b$;

$\therefore\quad M = 1\backslash a_b \ldots\ldots h\backslash t_b \ldots\ldots k\backslash s_b \ldots\ldots$, which is a Constituent of the new Block;

$\therefore$ for any Constituent $M$ in the Determinant of the original Block, there is a Constituent in that of the new Block, of the same absolute value;

and the symbol representing the one may be deduced from that representing the other by *one* interchange of consequents;

$\therefore$ the 2 symbols are of different classes; (Ch. I. Prop. IV.

$\therefore$ the Constituents, represented by them, have opposite signs;

$\therefore$ the whole Determinants are equal in value, but have opposite signs.

Similarly, if 2 columns be interchanged.

Therefore, if there be &c., Q. E. D*.

## Corollary to Prop. II.

If, in a square Block, a row, or a column, be made to pass over the next $r$ rows, or columns, either way: the Determinant of the new Block has the same sign as that of the first, or the opposite sign, according as $r$ is even or odd: that is, it is equal to the Determinant of the first Block multiplied by $(-1)^{r-1}$.

For this may be effected by interchanging it with each of these $r$ rows, or columns, in turn; and after *one* such interchange, the sign of the Determinant is changed, after *two*, it is the same again, and so on†.

---

* *Prop. II.* Thus the Determinant

$$\begin{vmatrix} a & b & c \\ d & e & f \\ g & h & l \end{vmatrix} = -\begin{vmatrix} c & b & a \\ f & e & d \\ l & h & g \end{vmatrix}.$$

† *Prop. II. Cor.* Thus the Determinant

$$\begin{vmatrix} a & b & c \\ d & e & f \\ g & h & k \end{vmatrix} = -\begin{vmatrix} b & a & c \\ e & d & f \\ h & g & k \end{vmatrix},$$

where the first column has been passed over *one* column: but the same Determinant

$$= +\begin{vmatrix} b & c & a \\ e & f & d \\ h & k & g \end{vmatrix},$$

where the first column has been passed over *two* columns.

## PROPOSITION III. TH.

If, in a square Block, 2 rows, or 2 columns, be identical: the Determinant vanishes.

Call the Determinant "$D$".

Now if the 2 identical rows, or columns, be interchanged, the Determinant of the new Block $= -D$; (PROP. II.

but the new Block is identical with the first;

$\therefore \quad D = -D$;

i. e. $D = 0$.

Therefore, if in a square Block, &c. Q. E. D *.

## COROLLARIES TO PROP. III.

1.

If a square Block of the $n^{\text{th}}$ degree be represented by

$$\left\{\begin{matrix} 1\backslash 1 \cdots\cdots 1\backslash n \\ \vdots \qquad\quad \vdots \\ n\backslash 1 \cdots\cdots n\backslash n \end{matrix}\right\}_a,$$

and its adjugate Block by

$$\left\{\begin{matrix} 1\backslash 1 \cdots\cdots 1\backslash n \\ \vdots \qquad\quad \vdots \\ n\backslash 1 \cdots\cdots n\backslash n \end{matrix}\right\}_A,$$

then $$r\backslash 1_a . s\backslash 1_A + r\backslash 2_a . s\backslash 2_A + \cdots\cdots + r\backslash n_a . s\backslash n_A = 0,$$

and $$1\backslash r_a . 1\backslash s_A + 2\backslash r_a . 2\backslash s_A + \cdots\cdots + n\backslash r_a . n\backslash s_A = 0,$$

so long as $r \neq s$. For the quantity $s\backslash 1_a . s\backslash 1_A +$&c. is the Determinant of

---

* *Prop. III.* Thus the Determinant $\begin{vmatrix} a & b & c \\ a & b & c \\ d & e & f \end{vmatrix} = 0$.

the given Block, and if, for the several terms of the $s^{\text{th}}$ row, there be substituted terms equal to those in the $r^{\text{th}}$ row, this may be written $r\backslash 1_{a} . s\backslash 1 ._{A} + \&c.$; and since the new Block, so formed, has 2 rows identical, its Determinant vanishes.

2.

If, in a Block whose length exceeds its breadth by unity, the Elements of any one longitudinal be each multiplied by the Determinant of the Minor formed by erasing the lateral containing that Element: the sum of these products, affected with $+$ and $-$ alternately, is zero *.

3.

If, in a square Block, there be added to the several Elements of any row, or column, the corresponding Elements of any other row, or column, multiplied by any number: the Determinant of the new Block is the same as that of the first †.

## Proposition IV. Th.

If there be a square Block, and if, retaining the first term of the first row in its place, the rows be made columns, and the columns rows: the Determinant of the new Block is equal to that of the first.

---

* *Prop. III. Cor.* 2. Thus, in the Block

$$\left\{\begin{matrix} a & b & c & d \\ e & f & g & h \\ k & l & m & n \end{matrix}\right\},$$

the sum of

$$a.\begin{vmatrix} b & c & d \\ f & g & h \\ l & m & n \end{vmatrix} - b.\begin{vmatrix} a & c & d \\ e & g & h \\ k & m & n \end{vmatrix} + c\begin{vmatrix} a & b & d \\ e & f & h \\ k & l & n \end{vmatrix} - d\begin{vmatrix} a & b & c \\ e & f & g \\ k & l & m \end{vmatrix}$$

is zero: for this is the same thing as the Determinant of the Block $\left\{\begin{matrix} a & b & c & d \\ a & b & c & d \\ e & f & g & h \\ k & l & m & h \end{matrix}\right\}$, and, as this Block has 2 identical rows, its Determinant must vanish.

† *Prop. III. Cor.* 3. Thus, in the Block

$$\left\{\begin{matrix} a & b & c \\ d & e & f \\ g & h & k \end{matrix}\right\},$$

let us add to the first column the Elements of the third, multiplied by $m$: then the Determinant of the new Block is $\begin{vmatrix} (a+mc), & b, & c \\ (d+mf), & e, & f \\ (g+mk), & h, & k \end{vmatrix}$, and this is equal to

$$\begin{vmatrix} a & b & c \\ d & e & f \\ g & h & k \end{vmatrix} + m\begin{vmatrix} c & b & c \\ f & e & f \\ k & h & k \end{vmatrix} = \begin{vmatrix} a & b & c \\ d & e & f \\ g & h & k \end{vmatrix},$$

since the second Determinant vanishes.

Let the Block be represented by

$$\left\{\begin{matrix} 1\backslash 1 \ldots\ldots 1\backslash n \\ \vdots \qquad \vdots \\ n\backslash 1 \ldots\ldots n\backslash n \end{matrix}\right\}_a$$

and the new Block by

$$\left\{\begin{matrix} 1\backslash 1 \ldots\ldots 1\backslash n \\ \vdots \qquad \vdots \\ n\backslash 1 \ldots\ldots n\backslash n \end{matrix}\right\}_b;$$

and let a certain Constituent of the original Block, arranged in order of antecedents, be represented by $1\backslash\alpha_a . 2\backslash\beta_a \ldots\ldots n\backslash\zeta_a$, where $\alpha, \beta, \ldots\ldots \zeta$, are a certain permutation of the numbers 1 to $n$;

now $1\backslash\alpha_a = \alpha\backslash 1_b$, $2\backslash\beta_a = \beta\backslash 2_b, \ldots\ldots n\backslash\zeta_a = \zeta\backslash n_b$;

hence this Constituent $= \alpha\backslash 1_b . \beta\backslash 2_b \ldots\ldots \zeta\backslash n_b$;

that is, it = a certain Constituent of the new Block, arranged in order of consequents;

and the antecedents in the second case are the same permutation of the numbers 1 to $n$ as the consequents in the first case;

∴ the two Constituents are of the same class; (CHAP. I. DEF. III.

∴ they have the same sign; (CHAP. II. DEF. VIII.

∴ for every Constituent of the original Block, there is one of the new Block, equal to it, and with the same sign;

∴ the two Determinants are equal.

Therefore, if there be, &c. Q. E. D *.

---

* *Prop. IV.* Thus the Determinant

$$\begin{vmatrix} a & b & c \\ d & e & f \\ g & h & k \end{vmatrix} = \begin{vmatrix} a & d & g \\ b & e & h \\ c & f & k \end{vmatrix}.$$

## Proposition V. Th.

If there be given 2 Blocks, each consisting of $n$ rows and $r$ columns; and if each row of one Block be combined with each row of the other, by the process of multiplying the first term of one by the first term of the other, the second by the second, and so on, and adding the products; and if the $n^2$ quantities, so formed, be arranged as a square Block, in such a way that the Elements of the first row of the new Block are all formed from the first row of the first Block, by combining it with the $n$ rows of the second Block successively, and so on; then

firstly,

if $r < n$: the Determinant of the new Block vanishes:

secondly,

if $r = n$: the Determinant of the new Block is the product of the Determinants of the 2 given Blocks:

thirdly,

if $r > n$: the Determinant of the new Block is the sum of all possible products that can be made, by taking any $n$ columns of one of the given Blocks, in the order in which they stand, and the corresponding $n$ columns of the other Block, and multiplying together the Determinants of the 2 Blocks so formed.

Let the 2 Blocks be represented by

$$\left\{\begin{matrix} 1\backslash 1 \ldots\ldots 1\backslash r \\ \vdots \qquad\quad \vdots \\ n\backslash 1 \ldots\ldots n\backslash r \end{matrix}\right\}_a, \text{ and } \left\{\begin{matrix} 1\backslash 1 \ldots\ldots 1\backslash r \\ \vdots \qquad\quad \vdots \\ n\backslash 1 \ldots\ldots n\backslash r \end{matrix}\right\}_b;$$

and the new Block by

$$\left\{\begin{matrix} 1\backslash 1 \ldots\ldots 1\backslash n \\ \vdots \qquad\quad \vdots \\ n\backslash 1 \ldots\ldots n\backslash n \end{matrix}\right\}_c,$$

wherein any Element $h\backslash k_c = \{h\backslash 1 \ldots\ldots h\backslash r\}_a \S \{k\backslash 1 \ldots\ldots k\backslash r\}_b$;

$$= \Sigma\{h\backslash a_a . k\backslash a_b\},$$

in which $a$ takes all values from 1 to $r$;

now let a certain Constituent of the Determinant of the new Block be arranged in order of antecedents, and be represented by $1\backslash Q_c . 2\backslash R_c \ldots\ldots n\backslash T_c$, in which $Q, R, \ldots\ldots T$, are a certain permutation of the numbers 1 to $n$;

then this Constituent

$$= \Sigma\{1\backslash \alpha_a . Q\backslash \alpha_b\} . \Sigma\{2\backslash \beta_a . R\backslash \beta_b\} \ldots\ldots \Sigma\{n\backslash \delta_a . T\backslash \delta_b\};$$

in which *each* of the quantities $\alpha, \beta, \ldots\ldots \delta$, takes all values from 1 to $r$;

$$\therefore \text{ it } = \Sigma\{1\backslash \alpha_a . Q\backslash \alpha_b . 2\backslash \beta_a . R\backslash \beta_b \ldots\ldots n\backslash \delta_a . T\backslash \delta_b\};$$

$$= \Sigma\{1\backslash \alpha_a . 2\backslash \beta_a \ldots\ldots n\backslash \delta_a . Q\backslash \alpha_b . R\backslash \beta_b \ldots\ldots T\backslash \delta_b\};$$

also this Constituent is affected with + or −, according as the series $Q, R, \ldots\ldots T$, contains an even or odd number of derangements;

∴ the Determinant of the new Block

$$= \Sigma\{1\backslash \alpha_a . 2\backslash \beta_a \ldots\ldots n\backslash \delta_a . Q\backslash \alpha_b . R\backslash \beta_b \ldots\ldots T\backslash \delta_b\},$$

in which not only does each of the quantities $\alpha, \beta, \ldots\ldots \delta$, take all values from 1 to $r$, but also the series $Q, R, \ldots\ldots T$, takes the values of all possible permutations of the numbers 1 to $n$;

$$\therefore \text{ it } = \Sigma\{1\backslash \alpha_a . 2\backslash \beta_a \ldots\ldots n\backslash \delta_a . \Sigma(Q\backslash \alpha_b . R\backslash \beta_b \ldots\ldots T\backslash \delta_b)\};$$

wherein, whatsoever values are assigned to $\alpha, \beta, \ldots\ldots \delta$, in the outer bracket, the same are assigned to them in the inner bracket:

now the sum $\Sigma(Q\backslash \alpha_b . R\backslash \beta_b \ldots\ldots T\backslash \delta_b)$, each term of which is affected with + or −, according as the series $Q, R, \ldots\ldots T$, contains an even or odd number of derangements, is the Determinant

$$\begin{vmatrix} 1\backslash \alpha \ldots\ldots 1\backslash \delta \\ \vdots \qquad\qquad \vdots \\ n\backslash \alpha \ldots\ldots n\backslash \delta \end{vmatrix}_b;$$

that is, it is the Determinant of the square Block formed by taking from the $b$-Block its $\alpha^{\text{th}}$ column, its $\beta^{\text{th}}$ column, and so on, until $n$ columns have been taken, it being immaterial whether these be all different, or one or more of them be repeated any number of times.

∴ the Determinant of the new Block

$$= \Sigma \left\{ 1\backslash\alpha_a . 2\backslash\beta_a \ldots\ldots n\backslash\delta_a . \begin{vmatrix} 1\backslash\alpha \ldots\ldots 1\backslash\delta \\ \vdots \qquad\quad \vdots \\ n\backslash\alpha \ldots\ldots n\backslash\delta \end{vmatrix}_b \right\}.$$

Now, firstly,

let $r$ be $< n$;

then it is not possible to take from the $b$-Block $n$ different columns;

∴ the Determinant

$$\begin{vmatrix} 1\backslash\alpha \ldots\ldots 1\backslash\delta \\ \vdots \qquad\quad \vdots \\ n\backslash\alpha \ldots\ldots n\backslash\delta \end{vmatrix}_b$$

always contains 2 identical columns;

∴ it always vanishes; (PROP. III.

∴ the Determinant of the new Block vanishes.

Secondly,

let $r = n$;

then, if the series $\alpha, \beta, \ldots\ldots \delta$, be a permutation of the numbers 1 to $n$, the Determinant

$$\begin{vmatrix} 1\backslash\alpha \ldots\ldots 1\backslash\delta \\ \vdots \qquad\quad \vdots \\ n\backslash\alpha \ldots\ldots n\backslash\delta \end{vmatrix}_b = \begin{vmatrix} 1\backslash 1 \ldots\ldots 1\backslash n \\ \vdots \qquad\quad \vdots \\ n\backslash 1 \ldots\ldots n\backslash n \end{vmatrix}_b,$$

affected with +, or −, according as the series $\alpha, \beta, \ldots\ldots \delta$, contains an even or odd number of derangements; for either of these may be obtained from the other by interchanging columns; hence the 2 Determinants have the same absolute magnitudes, and have the same sign, or not, according as their diagonals have the same sign or not;

but if the series $\alpha, \beta, \ldots\ldots \delta$, be not such a permutation, the Determinant

$$\begin{vmatrix} 1\backslash\alpha \ldots\ldots 1\backslash\delta \\ \vdots \qquad\quad \vdots \\ n\backslash\alpha \ldots\ldots n\backslash\delta \end{vmatrix}_b$$

vanishes as in the first case;

∴ the Determinant of the new Block

$$= \Sigma\left\{\pm 1\backslash\alpha_a . 2\backslash\beta_a \ldots\ldots n\backslash\delta_a . \begin{vmatrix} 1\backslash 1 & \ldots\ldots & 1\backslash n \\ \vdots & & \vdots \\ n\backslash 1 & \ldots\ldots & n\backslash n \end{vmatrix}_b \right\};$$

$$= \Sigma\left\{\pm 1\backslash\alpha_a . 2\backslash\beta_a \ldots\ldots n\backslash\delta_a\right\} . \begin{vmatrix} 1\backslash 1 & \ldots\ldots & 1\backslash n \\ \vdots & & \vdots \\ n\backslash 1 & \ldots\ldots & n\backslash n \end{vmatrix}_b ;$$

$$= \begin{vmatrix} 1\backslash 1 & \ldots\ldots & 1\backslash n \\ \vdots & & \vdots \\ n\backslash 1 & \ldots\ldots & n\backslash n \end{vmatrix}_a \times \begin{vmatrix} 1\backslash 1 & \ldots\ldots & 1\backslash n \\ \vdots & & \vdots \\ n\backslash 1 & \ldots\ldots & n\backslash n \end{vmatrix}_b ;$$

that is, it is the product of the Determinants of the 2 given Blocks *.

Thirdly,

let $r$ be $> n$;

and let $A, B, \ldots\ldots N$, be a certain set of $n$ different numbers, selected from the numbers 1 to $r$, and orderly arranged; and let $\alpha, \beta, \ldots\ldots \delta$, *each* take any of the values $A, B, \ldots\ldots N$;

then, if the series $\alpha, \beta, \ldots\ldots \delta$, be a permutation of the numbers $A, B, \ldots\ldots N$,

$$\text{the Determinant } \begin{vmatrix} 1\backslash\alpha & \ldots\ldots & 1\backslash\delta \\ \vdots & & \vdots \\ n\backslash\alpha & \ldots\ldots & n\backslash\delta \end{vmatrix}_b = \begin{vmatrix} 1\backslash A & \ldots\ldots & 1\backslash N \\ \vdots & & \vdots \\ n\backslash A & \ldots\ldots & n\backslash N \end{vmatrix}_b ,$$

affected with + or −, according as the series $\alpha, \beta, \ldots\ldots \delta$, contains an even or odd number of derangements;

but if the series $\alpha, \beta, \ldots\ldots \delta$, be not such a permutation, the Determinant

---

* *Prop. V. Part* 2. Thus

$$\begin{vmatrix} a & b & c \\ d & e & f \\ g & h & k \end{vmatrix} \times \begin{vmatrix} A & B & C \\ D & E & F \\ G & H & K \end{vmatrix} = \begin{vmatrix} (aA+bB+cC), & (aD+bE+cF), & (aG+bH+cK) \\ (dA+eB+fC), & (dD+eE+fF), & (dG+eH+fK) \\ (gA+hB+kC), & (gD+hE+kF), & (gG+hH+kK) \end{vmatrix} .$$

$$\begin{vmatrix} 1\setminus a \ldots\ldots 1\setminus \delta \\ \vdots \qquad\quad \vdots \\ n\setminus a \ldots\ldots n\setminus \delta \end{vmatrix}_b$$

vanishes as in the first case;

∴ the Determinant of the new Block contains the quantity

$$\Sigma \left\{ \pm 1\setminus a_a . 2\setminus \beta_a \ldots\ldots n\setminus \delta_a . \begin{vmatrix} 1\setminus A \ldots\ldots 1\setminus N \\ \vdots \qquad\quad \vdots \\ n\setminus A \ldots\ldots n\setminus N \end{vmatrix}_b \right\};$$

∴ it contains $\Sigma \left\{ \pm 1\setminus a_a . 2\setminus \beta_a \ldots\ldots n\setminus \delta_a \right\} \times \begin{vmatrix} 1\setminus A \ldots\ldots 1\setminus N \\ \vdots \qquad\quad \vdots \\ n\setminus A \ldots\ldots n\setminus N \end{vmatrix}_b$;

∴ it contains $\begin{vmatrix} 1\setminus A \ldots\ldots 1\setminus N \\ \vdots \qquad\quad \vdots \\ n\setminus A \ldots\ldots n\setminus N \end{vmatrix}_a \times \begin{vmatrix} 1\setminus A \ldots\ldots 1\setminus N \\ \vdots \qquad\quad \vdots \\ n\setminus A \ldots\ldots n\setminus N \end{vmatrix}_b$;

and the same thing may be proved for any other set of $n$ different numbers, selected from the numbers 1 to $r$, and orderly arranged;

but if the series $A, B, \ldots\ldots N$, though selected from the numbers 1 to $r$, be not all *different* numbers, the Determinant

$$\begin{vmatrix} 1\setminus A \ldots\ldots 1\setminus N \\ \vdots \qquad\quad \vdots \\ n\setminus A \ldots\ldots n\setminus N \end{vmatrix}$$

vanishes as in the first case;

∴ the Determinant of the new Block

$$= \Sigma \left\{ \begin{vmatrix} 1\setminus A \ldots\ldots 1\setminus N \\ \vdots \qquad\quad \vdots \\ n\setminus A \ldots\ldots n\setminus N \end{vmatrix}_a \times \begin{vmatrix} 1\setminus A \ldots\ldots 1\setminus N \\ \vdots \qquad\quad \vdots \\ n\setminus A \ldots\ldots n\setminus N \end{vmatrix}_b \right\},$$

in which the series $A, \ldots\ldots N$, take the values of every possible set of $n$ different numbers, selected from the numbers 1 to $r$, and orderly arranged;

that is, it is the sum of all possible products that can be made, by taking any $n$ columns of one of the given Blocks, in the order in which they stand,

and the corresponding $n$ columns of the other Block, and multiplying together the Determinants of the 2 Blocks so formed.

Therefore, if there be given 2 Blocks, &c. Q. E. D.

## COROLLARY TO PROP. V.

If $r = n$: then, in each of the given Blocks, rows may be made columns, and columns rows, without altering the Determinants; (PROP. IV.

$\therefore$ the new Block may be such that any Element of it, $h\backslash k_c$, has any one of the 4 values,

$$\{h\backslash 1 \ldots\ldots h\backslash n\}_a \,\S\, \{k\backslash 1 \ldots\ldots k\backslash n\}_b,$$

$$\{h\backslash 1 \ldots\ldots h\backslash n\}_a \,\S\, \{1\backslash k \ldots\ldots n\backslash k\}_b,$$

$$\{1\backslash h \ldots\ldots n\backslash h\}_a \,\S\, \{k\backslash 1 \ldots\ldots k\backslash n\}_b,$$

$$\{1\backslash h \ldots\ldots n\backslash h\}_a \,\S\, \{1\backslash k \ldots\ldots n\backslash k\}_b.$$

## PROPOSITION VI. TH.

If there be a square Block of the $n^{th}$ degree: the Determinant of the adjugate Block is equal to the $\overline{n-1}^{th}$ power of the Determinant of the first Block.

Let the Block be represented by

$$\begin{Bmatrix} 1\backslash 1 \ldots\ldots 1\backslash n \\ \vdots \qquad \vdots \\ n\backslash 1 \ldots\ldots n\backslash n \end{Bmatrix}_a, \text{ and the adjugate Block by } \begin{Bmatrix} 1\backslash 1 \ldots\ldots 1\backslash n \\ \vdots \qquad \vdots \\ n\backslash 1 \ldots\ldots n\backslash n \end{Bmatrix}_A;$$

and let a Block of the $n^{th}$ degree be formed, represented by

$$\begin{Bmatrix} 1\backslash 1 \ldots\ldots 1\backslash n \\ \vdots \qquad \vdots \\ n\backslash 1 \ldots\ldots n\backslash n \end{Bmatrix}_c,$$

and such that any term $h\backslash k_c = (h\backslash 1_a \ldots\ldots h\backslash n_a) \,\S\, (k\backslash 1_A \ldots\ldots k\backslash n_A)$;

and let their Determinants be represented by $D_a$, $D_A$, $D_c$.

Then $D_c = D_a . D_A$; (PROP. V.

now $h\backslash k_c = h\backslash 1_a . k\backslash 1_A + \ldots\ldots + h\backslash n_a . k\backslash n_A$;

$\therefore$ when $h = k$, $h\backslash k_c = D_a$; (AX. II.

and when $h \neq k$, $h\backslash k_c = 0$; (PROP. III. COR. 1.

$\therefore$ all the Elements of $D_c$ vanish, except $1\backslash 1_c, \ldots\ldots n\backslash n_c$, each of which $= D_a$;

$\therefore \quad D_c = D_a^n$;

i. e. $D_a . D_A = D_a^n$;

$\therefore \quad D_A = D_a^{n-1}$. Q. E. D *.

## PROPOSITION VII. TH.

If there be a square Block of the $n^{\text{th}}$ degree, and if in it any Minor of the $m^{\text{th}}$ degree be selected: the Determinant of the corresponding Minor in the adjugate Block is equal, in absolute magnitude, to the product of the $\overline{m-1}^{\text{th}}$ power of the Determinant of the first Block, multiplied by the Determinant of the Minor complemental to the one selected.

Also, if the numerals, indicating the selected rows, be represented by $\alpha, \beta, \ldots\ldots$, and those indicating the selected columns by $\kappa, \lambda, \ldots\ldots$; and their respective sums by $\Sigma(\alpha)$, $\Sigma(\kappa)$: the relationship of sign between the equal magnitudes will be secured by multiplying either of them by $(-1)^{m.(\Sigma(\alpha)+\Sigma(\kappa))}$.

---

* *Prop. VI.* Thus, if the first Block be $\left\{\begin{matrix} a & b & c \\ d & e & f \\ g & h & k \end{matrix}\right\}$; then

$$\left|\begin{matrix} \begin{vmatrix} e & f \\ h & k \end{vmatrix}, & -\begin{vmatrix} d & f \\ g & k \end{vmatrix}, & \begin{vmatrix} d & e \\ g & h \end{vmatrix} \\ -\begin{vmatrix} b & c \\ h & k \end{vmatrix}, & \begin{vmatrix} a & c \\ g & k \end{vmatrix}, & -\begin{vmatrix} a & b \\ g & h \end{vmatrix} \\ \begin{vmatrix} b & c \\ e & f \end{vmatrix}, & -\begin{vmatrix} a & c \\ d & f \end{vmatrix}, & \begin{vmatrix} a & e \\ d & e \end{vmatrix} \end{matrix}\right| = \begin{vmatrix} a & b & c \\ d & e & f \\ g & h & k \end{vmatrix}^2.$$

Let the Block be re-arranged, if necessary, by transposing rows and columns, so that the selected rows and columns shall stand first; and let it, when so arranged, be represented by

$$\left\{\begin{matrix} 1\backslash 1 \ldots\ldots 1\backslash n \\ \vdots \qquad\quad \vdots \\ n\backslash 1 \ldots\ldots n\backslash n \end{matrix}\right\}_a, \quad \text{and its adjugate Block by} \quad \left\{\begin{matrix} 1\backslash 1 \ldots\ldots 1\backslash n \\ \vdots \qquad\quad \vdots \\ n\backslash 1 \ldots\ldots n\backslash n \end{matrix}\right\}_A;$$

also let a Block of the $n^{\text{th}}$ degree be formed, represented by

$$\left\{\begin{matrix} 1\backslash 1 \ldots\ldots 1\backslash n \\ \vdots \qquad\quad \vdots \\ n\backslash 1 \ldots\ldots n\backslash n \end{matrix}\right\}_b,$$

and such that its first $m$ rows are identical with those of the $A$-Block, the rest of its diagonal consists of units, and all its other Elements are zero;

$$\text{hence,} \quad \left\{\begin{matrix} 1\backslash 1 \ldots\ldots 1\backslash n \\ \vdots \qquad\quad \vdots \\ n\backslash 1 \ldots\ldots n\backslash n \end{matrix}\right\}_b \quad \text{is identical with}$$

$$\left\{\begin{matrix} 1\backslash 1_A \ldots\ldots 1\backslash m_A, & 1\backslash \overline{m+1}_A, \ldots\ldots 1\backslash n_A \\ \vdots \qquad\qquad \vdots & \vdots \qquad\qquad\qquad \vdots \\ m\backslash 1_A \ldots\ldots m\backslash m_A, & m\backslash \overline{m+1}_A, \ldots\ldots m\backslash n_A \\ 0 \ldots\ldots\ldots\ldots 0\;, & 1\;, 0 \ldots\ldots\ldots\ldots 0 \\ \vdots & \ddots \qquad\qquad \vdots \\ \vdots & \qquad\ddots \qquad 0 \\ 0 \ldots\ldots\ldots\ldots\ldots & \ldots\ldots\ldots\ldots 0\;, 1 \end{matrix}\right\};$$

$$\therefore \quad D_b = \begin{vmatrix} 1\backslash 1_A \ldots\ldots 1\backslash m_A \\ \vdots \qquad\qquad \vdots \\ m\backslash 1_A \ldots\ldots m\backslash m_A \end{vmatrix}; \qquad \text{(PROP. I. COR. 3.}$$

also let a Block of the $n^{\text{th}}$ degree be formed, represented by

$$\left\{\begin{matrix} 1\backslash 1 \ldots\ldots 1\backslash n \\ \vdots \qquad\quad \vdots \\ n\backslash 1 \ldots\ldots n\backslash n \end{matrix}\right\}_c ,$$

and such that any term $h\backslash k_c = \left(h\backslash 1_a \ldots\ldots h\backslash n_a\right) \;\S\; \left(k\backslash 1_b \ldots\ldots k\backslash n_b\right)$;

$\therefore\;\; D_a \cdot D_b = D_c.$ (PROP. V.

Now, for all values of $k \not> m$,

if $h = k$, $\;h\backslash k_c = D_a$; (AX. II.

if $h \neq k$, $\;h\backslash k_c = 0$; (PROP. III. COR. 1.

i. e. in the first $m$ columns of the $c$-Block, all the Elements vanish, except $1\backslash 1_c$, $2\backslash 2_c, \ldots\ldots m\backslash m_c$, each of which $= D_a$;

also, for all values of $k > m$,

$$h\backslash k_c = \left(h\backslash 1_a, \ldots\ldots h\backslash k_a, \ldots\ldots h\backslash n_a\right) \;\S\; (0, \ldots\ldots 1, \ldots\ldots 0),$$

since all the terms of the second series vanish, except $k\backslash k_b$, which $= 1$;

$\therefore$ for all values of $k > m$, $h\backslash k_c = h\backslash k_a$;

i. e. in the $\overline{m+1}^{\text{th}}$ and following columns of the $c$-Block, the Elements are identical with the corresponding Elements of the $a$-Block;

i. e. $\left\{\begin{matrix} 1\backslash 1 \ldots\ldots 1\backslash n \\ \vdots \qquad\quad \vdots \\ n\backslash 1 \ldots\ldots n\backslash n \end{matrix}\right\}_c$ is identical with

$$\left\{\begin{matrix} D_a, 0 \ldots\ldots 0, & 1\backslash \underline{m+1}_a & \ldots\ldots\ldots\ldots & 1\backslash n_a \\ 0 \;\ddots \qquad \vdots & \vdots & & \vdots \\ \vdots \qquad\;\; 0 & & & \\ \vdots \qquad D_a, & m\backslash \underline{m+1}_a & \ldots\ldots\ldots\ldots & m\backslash n_a \\ \vdots \qquad\;\; 0, & \overline{m+1}\backslash \underline{m+1}_a & \ldots\ldots & \overline{m+1}\backslash n_a \\ \vdots \qquad\;\; \vdots & \vdots & & \vdots \\ 0 \ldots\ldots\ldots 0, & n\backslash \underline{m+1}_a & \ldots\ldots\ldots & n\backslash n_a \end{matrix}\right\}$$

$$\therefore\quad D_c = D_a^{\,m}.\left|\begin{matrix} \overline{m+1}\backslash\underline{m+1}_a \ldots\ldots & \overline{m+1}\backslash\underline{n}_a \\ \vdots & \vdots \\ n\backslash\underline{m+1}_a \ldots\ldots & n\backslash\underline{n}_a \end{matrix}\right|;$$

$$\therefore\quad D_a.D_b = \text{do.}$$

$$\therefore\quad D_a.\left|\begin{matrix} 1\backslash\underline{1}_A \ldots\ldots & 1\backslash\underline{m}_A \\ \vdots & \vdots \\ m\backslash\underline{1}_A \ldots\ldots & m\backslash\underline{m}_A \end{matrix}\right| = \text{do.}$$

$$\text{i. e.}\quad \left|\begin{matrix} 1\backslash\underline{1}_A \ldots\ldots & 1\backslash\underline{m}_A \\ \vdots & \vdots \\ m\backslash\underline{1}_A \ldots\ldots & m\backslash\underline{m}_A \end{matrix}\right| = D_a^{\,m-1}.\left|\begin{matrix} \overline{m+1}\backslash\underline{m+1}_a \ldots\ldots & \overline{m+1}\backslash\underline{n}_a \\ \vdots & \vdots \\ n\backslash\underline{m+1}_a \ldots\ldots & n\backslash\underline{n}_a \end{matrix}\right|$$

Now let the Determinant of the original Block, before re-arrangement, be represented by '$D$'; the Determinant of the Minor, complemental to the one selected, by '$C$'; and if a Block be formed, adjugate to the original Block before re-arrangement, and if in it a Minor be taken corresponding to the one selected, let the Determinant of this Minor be represented by '$J$'.

Then, having regard to absolute magnitude only,

$$D_a = D;$$

$$\left|\begin{matrix} \overline{m+1}\backslash\underline{m+1}_a \ldots\ldots & \overline{m+1}\backslash\underline{n}_a \\ \vdots & \vdots \\ n\backslash\underline{m+1}_a \ldots\ldots & n\backslash\underline{n}_a \end{matrix}\right| = C;$$

$$\left|\begin{matrix} 1\backslash\underline{1}_A \ldots\ldots & 1\backslash\underline{m}_A \\ \vdots & \vdots \\ m\backslash\underline{1}_A \ldots\ldots & m\backslash\underline{m}_A \end{matrix}\right| = J;$$

$$\therefore\quad J = \pm\, D^{m-1}.C.$$

Therefore, if there be, &c. Q. E. D*.

Secondly, the relationship of sign between these equal magnitudes will be secured by multiplying either by $(-1)^{m.(\Sigma(a)+\Sigma(\kappa))}$.

Let the numeral, indicating the last of the selected rows, be represented by '$\zeta$'.

Now, firstly, $D_a$ is equal to $D$ in absolute magnitude; and it has the same, or a different, sign, according as the number of rows, over which other rows are transposed, together with the number of columns, over which other columns are transposed, is even or odd;

for each transposition over one row, or over one column, changes the sign of the Determinant; (PROP. II.

now the first of the selected rows, namely the $a^{\text{th}}$, is transposed over all the preceding $\overline{a-1}$ rows;

the second, namely the $\beta^{\text{th}}$, is transposed over all the preceding $\overline{\beta-1}$ rows, excepting the $a^{\text{th}}$ row itself; that is, it is transposed over $\overline{\beta-2}$ rows;

similarly the $\gamma^{\text{th}}$ row is transposed over $\overline{\gamma-3}$ rows, and so on; and finally the $\zeta^{\text{th}}$ is transposed over $\overline{\zeta-m}$ rows;

$\therefore$ the number of rows, over which other rows are transposed,

$$= a+\beta+\ldots\ldots+\zeta-(1+2+\ldots\ldots+m);$$

$$= \Sigma(a)-\frac{m.(m+1)}{2};$$

similarly, the number of columns, over which other columns are transposed,

$$= \Sigma(\kappa)-\frac{m.(m+1)}{2};$$

$\therefore$ their sum $= \Sigma(a)+\Sigma(\kappa)-m.(m+1)$;

but $m.(m+1)$ is necessarily even;

$\therefore$ $D_a$ has the same sign as $D$, or a different one, according as $(\Sigma(a)+\Sigma(\kappa))$ is even or odd.

---

* *Prop. VII. Part I.* Thus, if the Block be $\left\{\begin{matrix} a & b & c & d \\ e & f & g & h \\ k & l & m & n \\ p & q & r & s \end{matrix}\right\}$, and if the Minor $\left\{\begin{matrix} f & g \\ q & r \end{matrix}\right\}$ be selected;

$$\text{then} \begin{Vmatrix} \begin{vmatrix} a & c & d \\ k & m & n \\ p & r & s \end{vmatrix}, -\begin{vmatrix} a & b & d \\ k & l & n \\ p & q & s \end{vmatrix} \\ \\ \begin{vmatrix} a & c & d \\ e & g & h \\ k & m & n \end{vmatrix}, -\begin{vmatrix} a & b & d \\ e & f & h \\ e & l & n \end{vmatrix} \end{Vmatrix} = \pm \begin{vmatrix} a & b & c & d \\ e & f & g & h \\ k & l & m & n \\ p & q & r & s \end{vmatrix}^{(2-1)} \times \begin{vmatrix} a & d \\ k & n \end{vmatrix}.$$

$$\therefore \quad D_a = D.(-1)^{\Sigma(a)+\Sigma(\kappa)};$$

$$\therefore \quad D_a^{m-1} = D^{m-1}.(-1)^{(\Sigma(a)+\Sigma(\kappa)).(m-1)}.$$

Secondly, $\begin{vmatrix} \overline{m+1}\backslash \underline{m+1}_a \ldots\ldots \overline{m+1}\backslash \underline{n}_a \\ \vdots \qquad\qquad \vdots \\ n\backslash \underline{m+1}_a \ldots\ldots n\backslash \underline{n}_a \end{vmatrix}$ is equal to $C$ in absolute magnitude, and has the same sign;

for the two Determinants contain the same Elements, arranged in the same order.

Thirdly, $\begin{vmatrix} 1\backslash \underline{1}_A \ldots\ldots 1\backslash \underline{m}_A \\ \vdots \qquad\qquad \vdots \\ m\backslash \underline{1}_A \ldots\ldots m\backslash \underline{m}_A \end{vmatrix}$ is equal to $J$ in absolute magnitude, and it has the same, or a different, sign, according as $(\Sigma(a)+\Sigma(\kappa))$ is even or odd;

for the Elements of the first of these Determinants are the determinantal coefficients of the Elements of the selected Minor, after the re-arrangement of the Block; and those of the other are the same coefficients before the re-arrangement;

hence the Elements of the one Determinant have the same absolute magnitude as those of the other, and the same, or a different, sign, according as the Determinant of the Block has, after the re-arrangement, the same, or a different, sign;

$$\therefore \quad \begin{vmatrix} 1\backslash \underline{1}_A \ldots\ldots 1\backslash \underline{m}_A \\ \vdots \qquad\qquad \vdots \\ m\backslash \underline{1}_A \ldots\ldots m\backslash \underline{m}_A \end{vmatrix} = J.(-1)^{\Sigma(a)+\Sigma(\kappa)}.$$

Hence, substituting in the Equation already established, we get

$$J.(-1)^{\Sigma(a)+\Sigma(\kappa)} = D^{m-1}.C.(-1)^{(\Sigma(a)+\Sigma(\kappa)).(m-1)};$$

$$\therefore \quad J = D^{m-1}.C.(-1)^{m.(\Sigma(a)+\Sigma(\kappa))} *.$$

Therefore, if the numerals, &c. Q. E. D.

---

* The resulting Equation really is $J = D^{m-1}.C.(-1)^{(m-2).(\Sigma(a)+\Sigma(\kappa))}$, but the factor $(-1)^{-2(\Sigma(a)+\Sigma(\kappa))} = 1$, and so may be neglected.

## COROLLARY TO PROP. VII.

If $D_a = 0$, the Determinant of any Minor in the adjugate Block $= 0$, since it contains $D_a$ as a factor.

As a particular case of this, let $m = 2$, and let the selected Minor contain the Elements common to the $h^{th}$ and $k^{th}$ rows and the $r^{th}$ and $s^{th}$ columns;

$$\text{then} \begin{vmatrix} h\backslash r_A, & h\backslash s_A \\ k\backslash r_A, & k\backslash s_A \end{vmatrix} = 0;$$

$$\text{i. e. } h\backslash r_A \,.\, k\backslash s_A = k\backslash r_A \,.\, h\backslash s_A;$$

$$\therefore\ h\backslash r_A : h\backslash s_A :: k\backslash r_A : k\backslash s_A;$$

$\therefore$ the ratios $h\backslash 1_A : h\backslash 2_A : h\backslash 3_A :$ &c. $: h\backslash n_A$ are independent of $h$.

# CHAPTER III.

## *ANALYSIS OF EQUATIONS.*

N.B. THE EQUATIONS DISCUSSED IN THIS CHAPTER ARE OF THE FIRST DEGREE ONLY.

### DEFINITIONS.

I.

Any quantity that is not zero is called **actual.**

II.

In any Equation, or set of Equations, any Algebraical quantity, which has an actual coefficient, is said to be **contained actually** in it or them.

III.

In any Equation, or set of Equations, all Numbers, and all Algebraical quantities whose values are determinable independently of the Equations, are called **Constants** with reference to them: but if there be in them Algebraical quantities which may bear any values whatsoever that are consistent with the truth of the Equations, these are called **Variables** with reference to them.

IV.

In an Equation containing Variables, any set of values which can be assigned to the Variables, consistently with the truth of the Equations, is called a set of **values for the Variables.**

V.

If a set of Equations, containing Variables, be such that there is a set of finite values for the Variables, they are said to be **consistent**; if not, **inconsistent.**

VI.

If there be an Equation, or set of consistent Equations, containing Variables, and another such Equation; and if, whatsoever values for the Variables satisfy the first Equation or set of Equations, the same also satisfy that other Equation: that other Equation is said to be **dependent** on the rest.

VII.

If 2 Equations be each dependent on the other: they are said to be **identical.**

---

## Conventions.

I.

If a Block of terms be distinguished by prefixing a letter, as "the $A$-Block:" let $\|A\|$ represent "the Determinants of all its principal Minors," $|A|$ "the Determinant of one of its principal Minors," and (if the Block be square) let $A$ represent "the Determinant of the Block."

Thus, if a Block be square, these three symbols will bear the same meaning.

Also, in an oblong Block,

$\|A\|=0$ may be read "all the $A$-Determinants vanish," or "the $A$-Block is evanescent."

$|A|=0$ may be read "one of the $A$-Determinants vanishes."

$|A|\neq 0$ may be read "one of the $A$-Determinants does not vanish," or "the $A$-Block is not evanescent."

$\|A\|\neq 0$ may be read "none of the $A$-Determinants vanish."

F

II.

If there be a set of Equations containing Variables: let the Block formed of the coefficients of the Variables be called "the *V*-Block," and let the Block formed of these together with the constant terms be called "the *B*-Block."

III.

If a set of Equations be said to contain *n* Variables, and if, in any one of them, any Variable be not actually contained, let it be understood that, in forming the *V*-Block or *B*-Block, such Variable is introduced with a zero coefficient. And if any one of the Equations contain no actual constant term, let it be understood that, in forming the *B*-Block, a zero term is introduced as the constant term *.

## AXIOMS.

I.

If there be a set of homogeneous Equations containing Variables: they may be satisfied by assigning to each Variable the value zero.

II.

But if there be a set of values for the Variables, whereof one is actual: at least one other is actual also.

---

* *Conv. III.* Thus, in the Equations

$$3x+2y \quad -1=0,$$
$$x \quad +5z \quad =0,$$
$$x-y \quad +3=0,$$

the *V*-Block is $\begin{Bmatrix} 3, & 2, & 0 \\ 1, & 0, & 5 \\ 1, & -1, & 0 \end{Bmatrix}$, and the *B*-Block $\begin{Bmatrix} 3, & 2, & 0, & -1 \\ 1, & 0, & 5, & 0 \\ 1, & -1, & 0, & 3 \end{Bmatrix}$.

Again, if the Equations

$$x+3y+2 = 0,$$
$$x-\ y-1 = 0,$$
$$2x+\ y+3 = 0,$$

be said to contain *two* Variables, their *V*-Block is $\begin{Bmatrix} 1, & 3 \\ 1, & -1 \\ 2, & 1 \end{Bmatrix}$; but if they be said to contain *three*, it is $\begin{Bmatrix} 1, & 3, & 0 \\ 1, & -1, & 0 \\ 2, & 1, & 0 \end{Bmatrix}$.

### III.

If there be a set of homogeneous Equations containing Variables: whatsoever values for the Variables satisfy them, any equimultiples of these values also satisfy them.

### IV.

If there be 2 identical Equations, containing $m$ Variables: there are $\overline{m-1}$ of the Variables, to which arbitrary values may be assigned.

## SECTION I.

---

### *Consistency of Equations under given conditions of evanescence of their* V-*Blocks or* B-*Blocks.*

---

#### Proposition I. Th.

If there be $n$ Equations containing $n$ Variables, and if $V \neq 0$: the Equations are consistent, and there is only one set of values for the Variables.

First, let $n = 2$.

Let the Equations be represented by

$$1\backslash 1.x_1 + 1\backslash 2.x_2 + 1\backslash 3 = 0,$$

$$2\backslash 1.x_1 + 2\backslash 2.x_2 + 2\backslash 3 = 0;$$

and let the Determinants of the principal Minors of the $B$-Block, formed by successively erasing the columns containing the Variables and the column of constants, be represented by the symbols $D_1$, $D_2$, $V$;

then, if the first Equation be multiplied by $2\backslash 2$, and the second by $-1\backslash 2$, and the 2 Equations added together, the coefficient of $x_2$ in the resulting Equation will be zero;

$\therefore$ the resulting Equation is

$$x_1.\left(1\backslash 1.2\backslash 2-2\backslash 1.1\backslash 2\right)+\left(1\backslash 3.2\backslash 2-2\backslash 3.1\backslash 2\right)=0;$$

i. e. $$x_1.\begin{vmatrix}1\backslash 1, 1\backslash 2\\ 2\backslash 1, 2\backslash 2\end{vmatrix}+\begin{vmatrix}1\backslash 3, 1\backslash 2\\ 2\backslash 3, 2\backslash 2\end{vmatrix}=0;$$

i. e. $$x_1.\begin{vmatrix}1\backslash 1, 1\backslash 2\\ 2\backslash 1, 2\backslash 2\end{vmatrix}-\begin{vmatrix}1\backslash 2, 1\backslash 3\\ 2\backslash 2, 2\backslash 3\end{vmatrix}=0;$$ (CH. II. PROP. II. COR.

i. e. $x_1.V-D_1=0.$

By a similar process it may be proved that

$$-x_2.V-D_2=0;$$

and since, by hypothesis, $V\neq 0$, these Equations may be divided throughout by $V$, and written

$$x_1=\frac{D_1}{V}, \qquad -x_2=\frac{D_2}{V}.$$

Next, these values shall satisfy both Equations;
for let them be substituted in the first;
then its left-hand side will become $\frac{1}{V}.\left(1\backslash 1.D_1-1\backslash 2.D_2+1\backslash 3.V\right)$;
and this $=0$; (CH. II. PROP. III. COR. 2.
similarly it may be proved for the second Equation;
$\therefore$ the Equations are consistent.

Also it is evident that, whatsoever values there are for the Variables, they may be proved equal to $\frac{D_1}{V}, \frac{-D_2}{V}$;

$\therefore$ there is only one set of values for the Variables.

## Secondly, let $n>2$.

Let the equations be represented by

$$1\backslash 1.x_1+1\backslash 2.x_2+\ldots\ldots+1\backslash n.x_n+1\backslash \underline{n+1}=0,$$

$$2\backslash 1.x_1+2\backslash 2.x_2+\ldots\ldots+2\backslash n.x_n+2\backslash \underline{n+1}=0,$$

&c.

$$n\backslash 1.x_1+n\backslash 2.x_2+\ldots\ldots+n\backslash n.x_n+n\backslash \underline{n+1}=0;$$

and let the Determinants of the principal Minors of the $B$-Block be represented, as before, by $D_1, \ldots\ldots D_n, V$.

Now the Block

$$\begin{Bmatrix} 1\backslash 2 & \ldots\ldots & 1\backslash n \\ \vdots & & \vdots \\ n\backslash 2 & \ldots\ldots & n\backslash n \end{Bmatrix}$$

contains $n$ rows and $\overline{n-1}$ columns;

hence, if the Elements of any column in it be respectively multiplied by the Determinants of its principal Minors, affected with + and − alternately, the sum of the products is zero; (CH. II. PROP. III. COR. 2.

hence, if the $n$ Equations be respectively multiplied by these Determinants, thus affected, and added together, the coefficients of $x_2, \ldots\ldots x_n$, in the resulting Equation will all be zero;

∴ the resulting Equation is

$$x_1 . \left\{ 1\backslash 1 . \begin{vmatrix} 2\backslash 2 & \ldots\ldots\ldots & 2\backslash n \\ \vdots & & \vdots \\ n\backslash 2 & \ldots\ldots\ldots & n\backslash n \end{vmatrix} - 2\backslash 1 . \begin{vmatrix} 1\backslash 2 & \ldots\ldots\ldots & 1\backslash n \\ 3\backslash 2 & \ldots\ldots\ldots & 3\backslash n \\ \vdots & & \vdots \\ n\backslash 2 & \ldots\ldots\ldots & n\backslash n \end{vmatrix} + \&c. \right\}$$

$$+ \left\{ 1\backslash \underline{n+1} . \begin{vmatrix} 2\backslash 2 & \ldots\ldots\ldots & 2\backslash n \\ \vdots & & \vdots \\ n\backslash 2 & \ldots\ldots\ldots & n\backslash n \end{vmatrix} - 2\backslash \underline{n+1} . \begin{vmatrix} 1\backslash 2 & \ldots\ldots\ldots & 1\backslash n \\ 3\backslash 2 & \ldots\ldots\ldots & 3\backslash n \\ \vdots & & \vdots \\ n\backslash 2 & \ldots\ldots\ldots & n\backslash n \end{vmatrix} + \&c. \right\} = 0;$$

$$\text{i.e.} \quad x_1 . \begin{vmatrix} 1\backslash 1 & \ldots\ldots & 1\backslash n \\ \vdots & & \vdots \\ n\backslash 1 & \ldots\ldots & n\backslash n \end{vmatrix} + \begin{vmatrix} 1\backslash \underline{n+1}, & 1\backslash 2 & \ldots\ldots & 1\backslash n \\ \vdots & \vdots & & \vdots \\ n\backslash \underline{n+1}, & n\backslash 2 & \ldots\ldots & n\backslash n \end{vmatrix} = 0;$$

$$\text{i.e.} \quad x_1 . \begin{vmatrix} 1\backslash 1 & \ldots\ldots & 1\backslash n \\ \vdots & & \vdots \\ n\backslash 1 & \ldots\ldots & n\backslash n \end{vmatrix} (-)^{n-1} \begin{vmatrix} 1\backslash 2 & \ldots\ldots & 1\backslash \underline{n+1} \\ \vdots & & \vdots \\ n\backslash 2 & \ldots\ldots & n\backslash \underline{n+1} \end{vmatrix} = 0; \quad \text{(CH. II. PROP. II. COR.}$$

$$\text{i.e.} \quad x_1 . V \, (-)^{n-1} D_1 = 0.$$

By a similar process it may be proved that

$$-x_2 . V \, (-)^{n-1} D_2 = 0,$$

&c.;

and, dividing these Equations throughout by $V$,

$$x_1 = \frac{(-)^n D_1}{V}, \quad -x_2 = \frac{(-)^n D_2}{V}, \text{ \&c.};$$

and since, by hypothesis, $V \neq 0$, these values are all finite.

Next, these values shall satisfy all the Equations;
for let them be substituted in the Equation

$$\hat{h}\backslash 1.x_1 + \hat{h}\backslash 2.x_2 + \ldots\ldots + \hat{h}\backslash n.x_n + \hat{h}\backslash \underline{n+1} = 0;$$

then the left-hand side of this Equation will become

$$\frac{(-1)^n}{V}.\left\{\hat{h}\backslash 1.D_1 - \hat{h}\backslash 2.D_2 + \ldots\ldots (-)^{n-1}\hat{h}\backslash n.D_n (-)^n \hat{h}\backslash \underline{n+1}.V\right\};$$

and this $= 0$; (CH. II. PROP. III. COR. 2.

similarly it may be proved for any other of the Equations;
∴ the Equations are consistent.
Also it is evident that, whatsoever values there are for the Variables, they may be proved equal to $\frac{(-)^n.D_1}{V}$, &c.;

∴ there is only one set of values for the Variables.

Therefore, if there be, &c. Q. E. D.

## COROLLARY TO PROP. I.

The values for the Variables may be briefly exhibited thus;—

$$x_1 : -x_2 : \text{\&c.} : (-1)^n :: D_1 : D_2 : \text{\&c.} : V;$$

or thus; $$\frac{x_1}{D_1} = \frac{-x_2}{D_2} = \text{\&c.} = \frac{(-1)^n}{V} *;$$

wherein it is to be observed that, if any of the quantities $D_1, D_2, \ldots\ldots D_n$ be zero, the value of the corresponding Variable must also be zero.

---

* *Prop. I.* Thus, in the Equations $\left.\begin{matrix} 2x+3y+1=0 \\ 5x-\ y-2=0 \end{matrix}\right\}$, we have $\dfrac{x}{\begin{vmatrix} 3, & 1 \\ -1, & -2 \end{vmatrix}} = \dfrac{-y}{\begin{vmatrix} 2, & 1 \\ 5, & -2 \end{vmatrix}} = \dfrac{1}{\begin{vmatrix} 2, & 3 \\ 5, & -1 \end{vmatrix}}$;

i. e. $\dfrac{x}{-5} = \dfrac{-y}{-9} = \dfrac{1}{-17}$; ∴ $x = \dfrac{5}{17}$, and $y = -\dfrac{9}{17}$.

## Proposition II. Th.

If there be $n$ Equations containing $\overline{n+r}$ Variables, and if $|V| \neq 0$: the Equations are consistent, and, if any non-evanescent principal Minor of the $V$-Block be selected, the $r$ Variables, whose coefficients are not contained in it, may have arbitrary values assigned to them; and, for each such set of arbitrary values, there is only one value for each of the remaining Variables.

Since $|V| \neq 0$, the Equations must contain actually $n$ at least of the Variables.

Let a non-evanescent principal Minor of the $V$-Block be selected, and let arbitrary values be given to the $r$ Variables whose coefficients are not contained in it;

then there are $n$ Equations, containing $n$ Variables, and such that their $V$-Block does not vanish;

$\therefore$ they are consistent, &c. (Prop. I.

Therefore, if there be, &c. Q. E. D †.

## Corollary to Prop. II.

If the Equations be homogeneous, and if $r = 1$; then, in every set of values for the Variables, the values bear to each other one and the same set of ratios.

For if the Determinants of the principal Minors of the $B$-Block be represented by $D_1, \ldots\ldots D_{n+1}$, it may be proved, as in the last Proposition, that

$$\frac{x_1}{D_1} = \frac{-x_2}{D_2} = \&c. = \frac{(-)^n x_{n+1}}{D_{n+1}}.$$

---

Again, in the Equations

$$3x - y + z + 4 = 0,$$
$$x + 3y - 5 = 0,$$
$$2x + y - 3z + 3 = 0,$$

we have

$$\frac{x}{\begin{vmatrix} -1, & 1, & 4 \\ 3, & 0, & -5 \\ 1, & -3, & 3 \end{vmatrix}} = \frac{-y}{\begin{vmatrix} 3, & 1, & 4 \\ 1, & 0, & -5 \\ 2, & -3, & 3 \end{vmatrix}} = \frac{z}{\begin{vmatrix} 3, & -1, & 4 \\ 1, & 3, & -5 \\ 2, & 1, & 3 \end{vmatrix}} = \frac{-1}{\begin{vmatrix} 3, & -1, & 1 \\ 1, & 3, & 0 \\ 2, & 1, & -3 \end{vmatrix}};$$

i. e. $\frac{x}{-35} = \frac{y}{70} = \frac{z}{35} = \frac{1}{35}$;

$\therefore$ $x = -1$, $y = 2$, $z = 1$.

† *Prop. II.* Thus, in the Equations

$$x - y - 2z + v - 4 = 0,$$
$$3x + 2y + 4z - 2v + 3 = 0,$$

we have $\begin{vmatrix} 1, & -1 \\ 3, & 2 \end{vmatrix} \neq 0$: hence we may assign arbitrary values to $z$ and $v$; let us assign to them the values 0, 2; then we have

$$x = 1, \quad y = -1.$$

## PROPOSITION III. TH.

If there be $n$ Equations containing $\overline{n-1}$ Variables, and if $B \neq 0$; the Equations are inconsistent.

It is evident that the Equations cannot be all homogeneous, and that they must contain actually all the Variables.

First, let $n=2$.

Let the Equations be represented by

$$1\backslash 1.x + 1\backslash 2 = Q_1 = 0,$$

$$2\backslash 1.x + 2\backslash 2 = Q_2 = 0.$$

Now the quantity $Q_1.2\backslash 1 - Q_2.1\backslash 1$

$$= 1\backslash 1.x.2\backslash 1 + 1\backslash 2.2\backslash 1$$

$$- 2\backslash 1.x.1\backslash 1 - 2\backslash 2.1\backslash 1;$$

and, in this quantity, the first column $= 0$, and the other $= -B$;

$$\therefore \quad Q_1.2\backslash 1 - Q_2.1\backslash 1 \neq 0;$$

$\therefore$ the value for $x$, which makes $Q_1 = 0$, cannot also make $Q_2 = 0$.

Secondly, let $n > 2$.

Let the $n$ Equations be represented by

$$1\backslash 1.x_1 + \ldots\ldots + 1\backslash \underline{n-1}.x_{n-1} + 1\backslash n = Q_1 = 0,$$

&c.

$$n\backslash 1.x_1 + \ldots\ldots + n\backslash \underline{n-1}.x_{n-1} + n\backslash n = Q_n = 0;$$

also let the Determinants of the principal Minors of the Block

$$\left\{ \begin{matrix} 1\backslash 1 \ldots\ldots 1\backslash \underline{n-1} \\ \vdots \qquad\quad \vdots \\ n\backslash 1 \ldots\ldots n\backslash \underline{n-1} \end{matrix} \right\}$$

be represented by $H_1, \ldots\ldots H_n$.

Now the quantity $Q_1.H_1 - Q_2 H_2 + \ldots\ldots (-)^{n-1} Q_n.H_n$

$$= 1\backslash 1.x_1.H_1 + 1\backslash 2.x_2.H_1 + \ldots\ldots + 1\backslash n.H_1$$

$$-\left(2\backslash 1.x_1.H_2 + 2\backslash 2.x_2.H_2 + \ldots\ldots + 2\backslash n.H_2\right)$$

$$+ \&c.$$

$$(-)^{n-1}\left(n\backslash 1.x_1.H_n + n\backslash 2.x_2.H_n + \ldots\ldots + n\backslash n.H_n\right);$$

and, in this quantity, the first $\overline{n-1}$ columns vanish;

(Ch. II. Prop. III. Cor. 2.

and the last column is equal to $\pm B$;

$$\therefore \quad Q_1.H_1 - Q_2.H_2 + \ldots\ldots (-)^{n-1} Q_n.H_n = \pm B;$$

$$\therefore \quad \text{it} \neq 0;$$

$\therefore$ whatsoever values for the Variables make $Q_1, \ldots\ldots Q_{n-1}$, each $= 0$, these cannot also make $Q_n = 0$.

Therefore, if there be, &c. Q. E. D*.

## Corollary to Prop. III.

If there be $n$ Equations, not all homogeneous, containing $\overline{n-r}$ Variables, and if $|B| \neq 0$: the Equations are inconsistent.

For then there must be among them $\overline{n-r+1}$ Equations, not all homogeneous, containing $\overline{n-r}$ Variables, and such that the Determinant of their whole Block does not vanish.

## Proposition IV. Th.

If there be $n$ Equations containing $n$ Variables, and if $V = 0$, but $|B| \neq 0$: the Equations are inconsistent.

It is evident that the Equations cannot be all homogeneous, and that they must contain actually $\overline{n-1}$ at least of the Variables.

---

* *Prop. III.* Thus the Equations

$$x + y - 3 = 0,$$
$$2x + 3y - 7 = 0,$$
$$x - y + 2 = 0,$$

are inconsistent.

First, let them contain actually only $\overline{n-1}$ of the Variables; (whence $V$ must $=0$).

Then, by the last Proposition, they are inconsistent.

Secondly, let them contain actually all the $n$ Variables.

If possible, let there be a set of values for the Variables, and call them $a_1$, $a_2$, &c.;

then it may be proved, as in Proposition I, that

$$a_1 . V = (-)^n D_1, \quad -a_2 . V = (-)^n D_2, \text{ \&c.};$$

and, since $V=0$, these Equations become

$$0 = D_1 = D_2 = \text{\&c.};$$

$\therefore$ $\|B\| = 0$, which is contrary to the hypothesis.

Therefore, if there be, &c. Q. E. D.

## PROPOSITION V. TH.

If there be $n$ Equations containing $\overline{n+r}$ Variables, and if $\|V\| = 0$, but $|B| \neq 0$: the Equations are inconsistent.

It is evident that the Equations cannot be all homogeneous, and that they must contain actually $\overline{n-1}$ at least of the Variables.

First, let them contain actually only $\overline{n-1}$ of the Variables.

Then, by Proposition III, they are inconsistent.

Secondly, let them contain actually only $n$ of the Variables.

Then, by Proposition IV, they are inconsistent.

Thirdly, let them contain actually more than $n$ of the Variables.

Now, if possible, let them be consistent.

Since there is in the $B$-Block at least one principal Minor whose Determinant does not vanish, let the Variables, whose coefficients are contained in such a Minor, be placed first, and let the Equations, so arranged, be represented by

$$1\backslash\underline{1}.x_1 + \ldots\ldots + 1\backslash\underline{n-1}.x_{n-1} + 1\backslash\underline{n}.x_n + 1\backslash\underline{n+1}.x_{n+1} + \ldots$$
$$\ldots + 1\backslash\underline{n+r}.x_{n+r} + 1\backslash\underline{n+r+1} = 0,$$

&c.

$$n\backslash\underline{1}.x_1 + \ldots\ldots + n\backslash\underline{n-1}.x_{n-1} + n\backslash\underline{n}.x_n + n\backslash\underline{n+1}.x_{n+1} + \ldots$$
$$\ldots + n\backslash\underline{n+r}.x_{n+r} + n\backslash\underline{n+r+1} = 0\,;$$

so that $\begin{vmatrix} 1\backslash\underline{1}\ldots\ldots 1\backslash\underline{n-1}, & 1\backslash\underline{n+r+1} \\ \vdots & \vdots \\ n\backslash\underline{1}\ldots\ldots n\backslash\underline{n-1}, & n\backslash\underline{n+r+1} \end{vmatrix} \neq 0$; let it $= C$;

now let the values, belonging to the $r$ Variables $x_{n+1}, \ldots\ldots x_{n+r}$, be $a_{n+1}, \ldots\ldots a_{n+r}$;

then the Equations become

$$1\backslash\underline{1}.x_1 + \ldots + 1\backslash\underline{n-1}.x_{n-1} + 1\backslash\underline{n}.x_n + \left(1\backslash\underline{n+1}.a_{n+1} + \ldots + 1\backslash\underline{n+r+1}\right) = 0,$$

&c.

$$n\backslash\underline{1}.x_1 + \ldots + n\backslash\underline{n-1}.x_{n-1} + n\backslash\underline{n}.x_n + \left(n\backslash\underline{n+1}.a_{n+1} + \ldots + n\backslash\underline{n+r+1}\right) = 0\,;$$

and, in these Equations,

$$D_n = \begin{vmatrix} 1\backslash\underline{1}\ldots\ldots 1\backslash\underline{n-1}, & \left(1\backslash\underline{n+1}.a_{n+1} + \ldots\ldots + 1\backslash\underline{n+r+1}\right) \\ \vdots & \vdots \\ n\backslash\underline{1}\ldots\ldots n\backslash\underline{n-1}, & \left(n\backslash\underline{n+1}.a_{n+1} + \ldots\ldots + n\backslash\underline{n+r+1}\right) \end{vmatrix},$$

$$= \begin{vmatrix} 1\backslash\underline{1}\ldots\ldots 1\backslash\underline{n-1}, & 1\backslash\underline{n+1} \\ \vdots & \vdots \\ n\backslash\underline{1}\ldots\ldots n\backslash\underline{n-1}, & n\backslash\underline{n+1} \end{vmatrix} . a_{n+1} + \ldots$$

$$\ldots + \begin{vmatrix} 1\backslash\underline{1}\ldots\ldots 1\backslash\underline{n-1}, & 1\backslash\underline{n+r+1} \\ \vdots & \vdots \\ n\backslash\underline{1}\ldots\ldots n\backslash\underline{n-1}, & n\backslash\underline{n+r+1} \end{vmatrix}\,; \qquad \text{(Ch. II. Ax. II.}$$

and, since $\|V\|=0$, each of these Determinants vanishes, excepting the last, which $= C$;

$\therefore\ D_n = C$;

$\therefore$ there are $n$ Equations, containing $n$ Variables, and such that, in them, $V=0$, but $|B| \neq 0$;

$\therefore$ they are inconsistent. (PROP. IV.

Therefore, if there be, &c. Q. E. D.

## PROPOSITION VI. TH.

If there be 2 Equations containing Variables; and if $\|B\|=0$: the Equations are identical.

Let the 2 Equations contain $m$ Variables, and be represented by

$$1\backslash 1.x_1 + 1\backslash 2.x_2 + \ldots\ldots + 1\backslash m.x_m + 1\backslash\underline{m+1} = Q_1 = 0,$$

$$2\backslash 1.x_1 + 2\backslash 2.x_2 + \ldots\ldots + 2\backslash m.x_m + 2\backslash\underline{m+1} = Q_2 = 0;$$

so that $\left\| \begin{matrix} 1\backslash 1, \ldots\ldots 1\backslash\underline{m+1} \\ 2\backslash 1, \ldots\ldots 2\backslash\underline{m+1} \end{matrix} \right\| = 0.$

Since $\left| \begin{matrix} 1\backslash 1, 1\backslash 2 \\ 2\backslash 1, 2\backslash 2 \end{matrix} \right| = 0$;

$\therefore\ 1\backslash 1.2\backslash 2 = 2\backslash 1.1\backslash 2$;

$\therefore\ 1\backslash 1 : 2\backslash 1 :: 1\backslash 2 : 2\backslash 2 ::$ (by symmetry) &c. $:: 1\backslash\underline{m+1} : 2\backslash\underline{m+1}$;
$:: k : l$ (say);

$\therefore\ Q_1 : Q_2 :: k : l$;

$\therefore$ whatsoever values for the Variables make $Q_1 = 0$, these also make $Q_2 = 0$; and whatsoever values make $Q_2 = 0$, these also make $Q_1 = 0$.

$\therefore$ the Equations are identical.

Therefore, if there be, &c. Q. E. D.

## COROLLARY TO PROP. VI.

If there be $n$ Equations containing Variables; and if there be one of them such that, when it is taken along with each of the remaining Equations successively, each pair of Equations, so formed, has its $B$-Block evanescent: the $n$ Equations are identical.

## PROPOSITION VII. TH.

If there be 2 Equations containing Variables; and if $|B| \neq 0$: the Equations are not identical.

Let the Equations contain $m$ Variables.

Now, if possible, let them be identical;

$\therefore$ they are consistent, and there are, among the Variables, $\overline{m-1}$ to which arbitrary values may be assigned; (AX. IV.

but, if $|V| \neq 0$, there are only $\overline{m-2}$ Variables to which arbitrary values may be assigned; (PROP. II.

which is absurd;

and, if $\|V\| = 0$, the Equations are inconsistent; (PROPS. III, IV, V.

$\therefore$ in either case, they are not identical.

Therefore, if there be, &c. Q. E. D.

## PROPOSITION VIII. TH.

If there be $n$ Equations containing $\overline{n-1}$ Variables; and if there be among them $\overline{n-1}$ Equations, which have their $V$-Block not evanescent; and if $B=0$: the Equations are consistent; and there is only one set of values for the Variables; and the remaining Equation is dependent on these $\overline{n-1}$ Equations.

Let a set of $\overline{n-1}$ Equations, having their $V$-Block not evanescent, be placed first, and let the $n$ Equations, so arranged, be represented by

$$1\backslash 1.x_1 + \ldots\ldots + 1\backslash \underline{n-1}.x_{n-1} + 1\backslash n = Q_1 = 0,$$

&c.

$$n\backslash 1.x_1 + \ldots\ldots + n\backslash \underline{n-1}.x_{n-1} + n\backslash n = Q_n = 0;$$

so that the first $\overline{n-1}$ of these Equations are consistent; and there is only one set of values for the Variables. (PROP. I.

First, let the $n$ Equations be all homogeneous; (whence $B$ must $= 0$).

Then they may be satisfied by assigning to each Variable the value zero; (AX. I.

and these values satisfy the last Equation.

Secondly, let the $n$ Equations be not all homogeneous.

Let the Determinants of the principal Minors of the Block

$$\left\{\begin{array}{ccc} 1\backslash 1 & \ldots\ldots & 1\backslash \underline{n-1} \\ \vdots & & \vdots \\ n\backslash 1 & \ldots\ldots & n\backslash \underline{n-1} \end{array}\right\}$$

be represented by $H_1, \ldots\ldots H_n$; so that $H_n \neq 0$.

Then it may be proved, as in Proposition III, that

$$Q_1.H_1 - Q_2.H_2 + \ldots\ldots (-)^{n-1} Q_n.H_n = \pm B;$$

$$\therefore \text{ it } = 0;$$

$\therefore$ those values for the Variables, which make $Q_1, \ldots\ldots Q_{n-1}$, each $= 0$, the same also make $Q_n.H_n = 0$;

but $H_n \neq 0$;

$\therefore$ these values make $Q_n = 0$;

$\therefore$ the $n$ Equations are consistent, and the last is dependent on the others.

Therefore, if there be, &c. Q. E. D.

## Corollary to Prop. VIII.

If there be $n$ Equations containing $\overline{n-r}$ Variables; and if there be among them $\overline{n-r}$ Equations, which have their $V$-Block not evanescent; and if, when these $\overline{n-r}$ Equations are taken along with each of the remaining Equations successively, each set of $\overline{n-r+1}$ Equations, so formed, has its $B$-Block evanescent: the Equations are consistent; and there is only one set of values for the Variables; and the remaining Equations are dependent on these $\overline{n-1}$ Equations.

For then those values for the Variables, which satisfy such a set of $\overline{n-r}$ Equations, satisfy also each of the remaining Equations.

## Proposition IX. Th.

If there be $n$ Equations containing $n$ Variables; and if there be among them $n-1$ Equations, which have their $V$-Block not evanescent; and if $\|B\| = 0$: the Equations are consistent; and, if any non-evanescent principal Minor of the $V$-Block of these $n-1$ Equations be selected, the Variable, whose coefficients are not contained in it, may have an arbitrary value assigned to it; and, for each such arbitrary value, there is only one set of values for the other Variables; and the remaining Equation is dependent on these $\overline{n-1}$ Equations.

It is evident that the Equations must contain actually $\overline{n-1}$ at least of the Variables.

First, let them contain actually only $\overline{n-1}$ of the Variables.

Then the Equations are consistent, and there is only one set of values for the Variables. (PROP. VIII.

Also, an arbitrary value may be given to the Variable which is not actually contained in them.

Secondly, let them contain actually all the Variables.

Let a set of $\overline{n-1}$ Equations, such as satisfy the hypothesis, be selected; and let a non-evanescent principal Minor of their *V*-Block be selected; and let the remaining Equation be taken along with them; and let the $\overline{n-1}$ Variables, whose coefficients are contained in this Minor, be placed first in all the $n$ Equations, and let the $n$ Equations, so arranged, be represented by

$$1\backslash 1.x_1 + \ldots\ldots + 1\backslash\underline{n-1}.x_{n-1} \quad + 1\backslash n.x_n + 1\backslash\underline{n+1} = 0,$$

&c.

$$\overline{n-1}\backslash 1.x_1 + \ldots\ldots + \overline{n-1}\backslash\underline{n-1}.x_{n-1} + \overline{n-1}\backslash n.x_n + \overline{n-1}\backslash\underline{n+1} = 0,$$

$$n\backslash 1.x_1 + \ldots\ldots + n\backslash\underline{n-1}.x_{n-1} \quad + n\backslash n.x_n + n\backslash\underline{n+1} = 0.$$

Now let an arbitrary value be assigned to the Variable $x_n$; and call it "$a$";

then the Determinant of the *B*-Block of the $n$ Equations

$$= \begin{vmatrix} 1\backslash 1, \ldots 1\backslash\underline{n-1}, \left(1\backslash n.a + 1\backslash\underline{n+1}\right) \\ \vdots \qquad\qquad \vdots \qquad\qquad \vdots \\ n\backslash 1, \ldots n\backslash\underline{n-1}, \left(n\backslash n.a + n\backslash\underline{n+1}\right) \end{vmatrix};$$

$$= \begin{vmatrix} 1\backslash 1, \ldots\ldots 1\backslash n \\ \vdots \qquad\qquad \vdots \\ n\backslash 1 \ldots\ldots n\backslash n \end{vmatrix}.a + \begin{vmatrix} 1\backslash 1 \ldots 1\backslash\underline{n-1}, 1\backslash\underline{n+1} \\ \vdots \qquad\qquad \vdots \qquad \vdots \\ n\backslash 1 \ldots n\backslash\underline{n-1}, n\backslash\underline{n+1} \end{vmatrix} = 0, \text{ by hypothesis;}$$

$\therefore$ there are $n$ Equations, containing $\overline{n-1}$ Variables, and such that their $B$-Block is evanescent, but $\overline{n-1}$ of them have their $V$-Block not evanescent;

$\therefore$ they are consistent, and there is only one set of values for the Variables, and the last Equation is dependent on the others. (PROP. VIII.

Therefore, if there be, &c. Q. E. D.

## COROLLARY TO PROP. IX.

If there be $n$ Equations containing $\overline{n+r}$ Variables; and if there be among them $\overline{n-1}$ Equations, which have their $V$-Block not evanescent; and if $\|B\|=0$: the Equations are consistent; and, if any non-evanescent principal Minor of the $V$-Block of these $\overline{n-1}$ Equations be selected, the $\overline{r+1}$ Variables, whose coefficients are not contained in it, may have arbitrary values assigned to them; and, for each such set of arbitrary values, there is only one set of values for the other Variables; and the remaining Equation is dependent on these $\overline{n-1}$ Equations.

## PROPOSITION X. TH.

If there be $n$ Equations, containing $n$ Variables; and if there be among them $\overline{n-k}$ Equations, which have their $V$-Block not evanescent; and if, when these $\overline{n-k}$ Equations are taken along with each of the remaining Equations successively, each set of $\overline{n-k+1}$ Equations, so formed, has its $B$-Block evanescent (whence also $\|B\|=0$): the Equations are consistent; and, if any non-evanescent principal Minor of the $V$-Block of these $\overline{n-k}$ Equations be selected, the $k$ Variables, whose coefficients are not contained in it, may have arbitrary values assigned to them; and, for each such set of arbitrary values, there is only one set of values for the other Variables; and the remaining Equations are dependent on these $\overline{n-k}$ Equations.

It is evident that the Equations must contain actually $\overline{n-k}$ at least of the Variables.

First, let them contain actually only $\overline{n-k}$ of the Variables.

Then they are consistent; and there is only one set of values for the Variables; and the remaining $k$ Equations are dependent on these $\overline{n-k}$ Equations; (PROP. VIII. COR.

also arbitrary values may be given to the $k$ Variables which are not actually contained in the given Equations.

Secondly, let them contain actually more than $\overline{n-k}$ of the Variables.

Let a set of $\overline{n-k}$ Equations, such as satisfy the hypothesis, be selected; and let a non-evanescent principal Minor of their *V*-Block be selected; and let one of the remaining Equations be taken along with them; and let the $\overline{n-k}$ Variables, whose coefficients are contained in this Minor, be placed first in these $\overline{n-k+1}$ Equations; and let these $\overline{n-k+1}$ Equations, so arranged, be represented by

$$1\backslash 1.x_1 + \ldots + 1\backslash \underline{n-k}.x_{n-k} + 1\backslash \underline{n-k+1}.x_{n-k+1} + \ldots + 1\backslash n.x_n + 1\backslash \underline{n+1} = 0,$$

&c.

$$\overline{n-k}\backslash 1.x_1 + \ldots + \overline{n-k}\backslash \underline{n-k}.x_{n-k} + \overline{n-k}\backslash \underline{n-k+1}.x_{n-k+1} + \ldots$$
$$\ldots + \overline{n-k}\backslash n.x_n + \overline{n-k}\backslash \underline{n+1} = 0,$$

$$\overline{n-k+1}\backslash 1.x_1 + \ldots + \overline{n-k+1}\backslash \underline{n-k}.x_{n-k} + \overline{n-k+1}\backslash \underline{n-k+1}.x_{n-k+1} + \ldots$$
$$\ldots + \overline{n-k+1}\backslash n.x_n + \overline{n-k+1}\backslash \underline{n+1} = 0.$$

Now let arbitrary values be assigned to the Variables $x_{n-k+1}$, &c.;

∴ there are $\overline{n-k+1}$ Equations, containing $\overline{n-k}$ Variables, and there are among them $\overline{n-k}$ Equations, which have their *V*-Block not evanescent, and it may be proved, as in the last Proposition, that the Determinant of their *B*-Block is evanescent;

∴ they are consistent, and there is only one set of values for the Variables; and the $\overline{n-k+1}|^{\text{th}}$ Equation is dependent on these $\overline{n-k}$ Equations; (PROP. VIII.

also, if any other of the remaining Equations be substituted for this $\overline{n-k+1}|^{\text{th}}$ Equation, the same thing may be proved.

Therefore, if there be, &c. Q. E. D.

## COROLLARY TO PROP. X.

If there be $n$ Equations, containing $\overline{n+r}$ Variables; and if there be among them $\overline{n-k}$ Equations, which have their $V$-Block not evanescent; and if, when these $\overline{n-k}$ Equations are taken along with each of the remaining Equations successively, each set of $\overline{n-k+1}$ Equations, so formed, has its $B$-Block evanescent (whence also $\|B\|=0$): the Equations are consistent; and, if any non-evanescent principal Minor of the $V$-Block of these $\overline{n-k}$ Equations be selected, the $\overline{k+r}$ Variables, whose coefficients are not contained in it, may have arbitrary values assigned to them; and, for each such set of arbitrary values, there is only one set of values for the other Variables; and the remaining Equations are dependent on these $\overline{n-k}$ Equations.

## PROPOSITION XI. TH.

If there be $n$ homogeneous Equations, containing $n$ Variables; and if $B=0$: there is, for the Variables, a set of values of which 2 at least are actual. And, of the $n$ Equations, one at least is dependent on the rest.

For, if there be among them $\overline{n-1}$ Equations which have their $V$-Block not evanescent, there is one Variable to which an arbitrary value may be assigned; (PROP. IX.

let an actual value be assigned to this Variable;

then at least one other Variable has an actual value; (AX. II.

and the remaining Equation is dependent on these $\overline{n-1}$ Equations. (PROP. IX.

But, if every $\overline{n-1}$ of them have their $V$-Block evanescent, and if the greatest number of them, which have their $V$-Block not evanescent, be $\overline{n-k}$ (so that $k>1$), then there are $k$ Variables to which arbitrary values may be assigned; (PROP. X.

that is, there are 2 at least to which actual values may be assigned;

and the remaining $k$ Equations are dependent on these $\overline{n-k}$ Equations. (PROP. X.

Therefore, if there be, &c. Q. E. D.

## COROLLARY TO PROP. XI.

If there be $n$ homogeneous Equations, containing $\overline{n-r}$ Variables; and if $\|B\|=0$: there is, for the Variables, a set of values of which 2 at least are actual. And, of the $n$ Equations, $\overline{r+1}$ at least are dependent on the rest.

## PROPOSITION XII. TH.

If there be $n$ homogeneous Equations containing more than $n$ Variables: there is, for the Variables, a set of values, of which 2 at least are actual.

First, let there be one of the Equations such that, when it is taken along with each of the others successively, each pair of Equations, so formed, has its $V$-Block evanescent;

then the $n$ Equations are identical; (PROP. VI. COR.

$\therefore$ there are at least $n$ Variables, to which arbitrary values may be assigned; (AX. IV.

and one of these values may be actual;

$\therefore$ at least one other may have an actual value. (AX. II.

Secondly, let there be $k$ of the Equations, where $k$ is one of the numbers $2 \ldots \overline{n-1}$, which have their $V$-Block not evanescent, and are such that, when they are taken along with each of the others successively, the set of $\overline{k+1}$ Equations, so formed, has its $V$-Block evanescent;

then there are at least $\overline{n-k+1}$ Variables, to which arbitrary values may be assigned; (PROP. II.

that is, there are at least 2 such Variables;

and these values may be actual.

Thirdly, let the $n$ Equations have their $V$-Block not evanescent;

then there is at least one Variable, to which an arbitrary value may be assigned; (PROP. II.

and this value may be actual;

$\therefore$ at least one other may have an actual value. (AX. II.

Therefore, if there be, &c. Q. E. D.

# CHAPTER III. *(Continued.)*

## SECTION II.

*Properties of Equations under given conditions of consistency.*

### PROPOSITION XIII. TH.

If there be $n$ Equations containing $\overline{n-1}$ Variables; and if they be consistent: $B=0$.

For if not, let $B \neq 0$;
then they are inconsistent; (PROP. III.
which is contrary to the hypothesis.

Therefore, if there be, &c. Q. E. D.

### COROLLARY TO PROP. XIII.

If there be $n$ Equations containing $\overline{n-r}$ Variables; and if they be consistent: $\|B\|=0$.

### PROPOSITION XIV. TH.

If there be $n$ Equations containing $n$ Variables; and if they be consistent; and if $V=0$: then $\|B\|=0$.

For if not, let $|B| \neq 0$;
then the Equations are inconsistent; (PROP. IV.
which is contrary to the hypothesis.

Therefore, if there be, &c. Q. E. D.

## COROLLARY TO PROP. XIV.

If there be $n$ homogeneous Equations containing $\overline{n+1}$ Variables; and if there be, for the Variables, a set of values of which one is actual; and if, when that column of the $V$-Block, which contains such a Variable, is omitted, the remaining Block be evanescent: then the whole $V$-Block is evanescent.

For the Variable, whose coefficients are contained in that column, may be considered as constant, and the rest as Variables.

## PROPOSITION XV. TH.

If there be $n$ Equations containing $\overline{n+r}$ Variables; and if they be consistent; and if $\|V\| = 0$: then $\|B\| = 0$.

For if not, let $|B| \neq 0$;
then the Equations are inconsistent; (PROP. V.
which is contrary to the hypothesis.

Therefore, if there be, &c. Q. E. D.

## COROLLARY TO PROP. XV.

If there be $n$ homogeneous Equations containing $\overline{n+r}$ Variables; and if there be, for the Variables, a set of values of which one is actual; and if, when that column of the $V$-Block, which contains such a Variable, is omitted, the remaining Block be evanescent: then the whole $V$-Block is evanescent.

For the Variable, whose coefficients are contained in that column, may be considered as constant, and the rest as Variables.

## PROPOSITION XVI. TH.

If there be 2 Equations containing Variables; and if they be identical: $\|B\| = 0$.

For if not, let $|B| \neq 0$;
then the Equations are not identical; (PROP. VII.
which is contrary to the hypothesis.

Therefore, if there be, &c. Q. E. D.

## PROPOSITION XVII. TH.

If there be $n$ Equations containing $n$ Variables; and if they be consistent, and there be one Variable to which an arbitrary value may be assigned: $\|B\| = 0$.

Let the Variable, to which an arbitrary value may be assigned, be placed last, and let the Equations, so arranged, be represented by

$$\overline{1}\backslash 1.x_1 + \ldots\ldots + \overline{1}\backslash \underline{n-1}.x_{n-1} + \overline{1}\backslash n.x_n + \overline{1}\backslash \underline{n+1} = 0,$$

&c.

$$n\backslash 1.x_1 + \ldots\ldots + n\backslash \underline{n-1}.x_{n-1} + n\backslash n.x_n + n\backslash \underline{n+1} = 0;$$

and call the Determinants of the principal Minors of their $B$-Block, $D_1, \ldots D_n, V$.

Now, if possible, let $|B| \neq 0$.

First, if possible, let $D_n \neq 0$;

let $x_n$ have the arbitrary value zero assigned to it;

then there are $n$ Equations containing $\overline{n-1}$ Variables, and such that in them $B \neq 0$;

$\therefore$ they are inconsistent; (PROP. III.

which is contrary to the hypothesis.

$\therefore$ $D_n = 0$.

Secondly, if possible, let $V \neq 0$;

let $x_n$ have the arbitrary value, 1, assigned to it;

then there are $n$ Equations containing $\overline{n-1}$ Variables, and such that the Determinant of their $B$-Block

$$= \begin{vmatrix} \overline{1}\backslash 1 \ldots\ldots \overline{1}\backslash \underline{n-1}, & \left(\overline{1}\backslash n + \overline{1}\backslash \underline{n+1}\right) \\ \vdots \qquad\qquad \vdots & \vdots \\ n\backslash 1 \ldots\ldots n\backslash \underline{n-1}, & \left(n\backslash n + n\backslash \underline{n+1}\right) \end{vmatrix}$$

$$= \begin{vmatrix} \overline{1}\backslash 1 \ldots\ldots n\backslash n \\ \vdots \qquad \vdots \\ n\backslash 1 \ldots\ldots n\backslash n \end{vmatrix} + \begin{vmatrix} \overline{1}\backslash 1 \ldots\ldots \overline{1}\backslash \underline{n-1}, \overline{1}\backslash \underline{n+1} \\ \vdots \qquad\qquad \vdots \qquad \vdots \\ n\backslash 1 \ldots\ldots n\backslash \underline{n-1}, n\backslash \underline{n+1} \end{vmatrix};$$

$$= \quad V \quad + \quad 0 \quad = V;$$

$\therefore$ it $\neq 0$;
$\therefore$ the Equations are inconsistent; (PROP. III.
which is contrary to the hypothesis;
$\therefore$ $V = 0$;
$\therefore$ $\|B\| = 0$. (PROP. XIV.

Therefore, if there be, &c. Q. E. D.

## PROPOSITION XVIII. TH.

If there be $n$ Equations, containing $\overline{n+r}$ Variables; and if they be consistent, and there be $\overline{r+1}$ Variables to which arbitrary values may be assigned: $\|B\| = 0$.

Let the $\overline{r+1}$ Variables, to which arbitrary values may be assigned, be placed last, and let the Equations, so arranged, be represented by

$$\overline{1}\backslash 1.x_1 + \ldots\ldots + \overline{1}\backslash\underline{n-1}.x_{n-1} + \overline{1}\backslash n.x_n + \ldots\ldots + \overline{1}\backslash\underline{n+r}.x_{n+r} + \overline{1}\backslash\underline{n+r+1} = 0,$$

&c.

$$n\backslash 1.x_1 + \ldots\ldots + n\backslash\underline{n-1}.x_{n-1} + n\backslash n.x_n + \ldots\ldots + n\backslash\underline{n+r}.x_{n+r} + n\backslash\underline{n+r+1} = 0.$$

Now, if possible, let $|B| \neq 0$.

First, if possible, let $$\begin{vmatrix} \overline{1}\backslash 1 \ldots\ldots \overline{1}\backslash\underline{n-1}, & \overline{1}\backslash\underline{n+r+1} \\ \vdots \quad\quad \vdots & \vdots \\ n\backslash 1 \ldots\ldots n\backslash\underline{n-1}, & n\backslash\underline{n+r+1} \end{vmatrix} \neq 0;$$

let $x_n, \ldots\ldots x_{n+r}$ have the arbitrary value zero assigned to each of them;

then there are $n$ Equations containing $\overline{n-1}$ Variables, and such that, in them, $B \neq 0$;

$\therefore$ the Equations are inconsistent; (PROP. III.
which is contrary to the hypothesis;
$\therefore$ this Determinant $= 0$.

Secondly, if possible, let $$\begin{vmatrix} \overline{1}\backslash 1 \ldots\ldots \overline{1}\backslash\underline{n-1}, & \overline{1}\backslash\underline{n+k} \\ \vdots \quad\quad \vdots & \vdots \\ n\backslash 1 \ldots\ldots n\backslash\underline{n-1}, & n\backslash\underline{n+k} \end{vmatrix} \neq 0;$$

wherein $k$ is some one of the numbers $0, 1, \ldots r$;

let $x_{n+k}$ have the arbitrary value, 1, assigned to it; and let each other of the Variables $x_n, \ldots\ldots x_{n+r}$ have the arbitrary value zero assigned to it;

then there are $n$ Equations containing $\overline{n-1}$ Variables, and such that their $B$-Block

$$= \begin{vmatrix} 1\backslash 1 \ldots\ldots 1\backslash \underline{n-1}, \left(1\backslash \underline{n+k} + 1\backslash \underline{n+r+1}\right) \\ \vdots \qquad\qquad \vdots \qquad\qquad \vdots \\ n\backslash 1 \ldots\ldots n\backslash \underline{n-1}, \left(n\backslash \underline{n+k} + n\backslash \underline{n+r+1}\right) \end{vmatrix};$$

$$= \begin{vmatrix} 1\backslash 1 \ldots\ldots 1\backslash \underline{n-1}, 1\backslash \underline{n+k} \\ \vdots \qquad \vdots \qquad \vdots \\ n\backslash 1 \ldots\ldots n\backslash \underline{n-1}, n\backslash \underline{n+k} \end{vmatrix} + \begin{vmatrix} 1\backslash 1 \ldots\ldots 1\backslash \underline{n-1}, 1\backslash \underline{n+r+1} \\ \vdots \qquad \vdots \qquad \vdots \\ n\backslash 1 \ldots\ldots n\backslash \underline{n-1}, n\backslash \underline{n+r+1} \end{vmatrix};$$

$$= \begin{vmatrix} 1\backslash 1 \ldots\ldots 1\backslash \underline{n-1}, 1\backslash \underline{n+k} \\ \vdots \qquad \vdots \qquad \vdots \\ n\backslash 1 \ldots\ldots n\backslash \underline{n-1}, n\backslash \underline{n+k} \end{vmatrix} + 0;$$

$\therefore$ it $\neq 0$;

$\therefore$ the Equations are inconsistent; (PROP. III.

which is contrary to the hypothesis;

$\therefore$ the square Blocks, formed by taking the first $\overline{n-1}$ columns along with each of the next $\overline{r+1}$ columns successively, are all evanescent;

that is, all the principal Minors of the $V$-Block, containing its first $\overline{n-1}$ columns, are evanescent.

Thirdly, let $\begin{vmatrix} 1\backslash 1 \ldots\ldots 1\backslash \underline{n-2}, 1\backslash \underline{n+k}, 1\backslash \underline{n+r+1} \\ \vdots \qquad \vdots \qquad \vdots \qquad \vdots \\ n\backslash 1 \ldots\ldots n\backslash \underline{n-2}, n\backslash \underline{n+k}, n\backslash \underline{n+r+1} \end{vmatrix} \neq 0$;

wherein $k$ is some one of the numbers $0, 1, \ldots r$;

let each of the Variables $x_n, \ldots x_{n+r}$, except $x_{n+k}$, have the arbitrary value zero assigned to it; and let $x_{n-1}$ have a possible value assigned to it, and call this value "$a$";

then there are $n$ Equations containing $\overline{n-1}$ Variables, and such that their $B$-Block

$$= \begin{vmatrix} \overline{1}\diagdown 1 \ldots\ldots \overline{1}\diagdown\underline{n-2}, \overline{1}\diagdown\underline{n+k}, \left(\overline{1}\diagdown\underline{n-1}.a+\overline{1}\diagdown\underline{n+r+1}\right) \\ \vdots \qquad\qquad \vdots \qquad\qquad \vdots \qquad\qquad\qquad \vdots \\ n\diagdown 1 \ldots\ldots n\diagdown\underline{n-2}, n\diagdown\underline{n+k}, \left(n\diagdown\underline{n-1}.a+n\diagdown\underline{n+r+1}\right) \end{vmatrix};$$

$$= \begin{vmatrix} \overline{1}\diagdown 1 \ldots\ldots \overline{1}\diagdown\underline{n-2}, \overline{1}\diagdown\underline{n+k}, \overline{1}\diagdown\underline{n-1} \\ \vdots \qquad\qquad \vdots \qquad\qquad \vdots \qquad\qquad \vdots \\ n\diagdown 1 \ldots\ldots n\diagdown\underline{n-2}, n\diagdown\underline{n+k}, n\diagdown\underline{n-1} \end{vmatrix} . a$$

$$+ \begin{vmatrix} \overline{1}\diagdown 1 \ldots\ldots \overline{1}\diagdown\underline{n-2}, \overline{1}\diagdown\underline{n+k}, \overline{1}\diagdown\underline{n+r+1} \\ \vdots \qquad\qquad \vdots \qquad\qquad \vdots \qquad\qquad \vdots \\ n\diagdown 1 \ldots\ldots n\diagdown\underline{n-2}, n\diagdown\underline{n+k}, n\diagdown\underline{n+r+1} \end{vmatrix};$$

$$= \qquad 0 \qquad + \qquad \text{do.};$$

$\therefore$ it$\neq 0$;

$\therefore$ the Equations are inconsistent; (PROP. III.

which is contrary to the hypothesis;

$\therefore$ the square Blocks, formed by taking the first $\overline{n-2}$ columns, along with one of the columns from the $n^{\text{th}}$ to the $\overline{n+r}|^{\text{th}}$, and with the last column, are all evanescent;

and the same thing may be proved for any $\overline{n-2}$ of the first $\overline{n-1}$ columns;

$\therefore$ the square Blocks, formed by taking any $\overline{n-2}$ of the first $\overline{n-1}$ columns, along with any one of the next $\overline{r+1}$ columns, and with the last column, are all evanescent.

Fourthly, let $\begin{vmatrix} \overline{1}\diagdown 1 \ldots\ldots \overline{1}\diagdown\underline{n-2}, \overline{1}\diagdown\underline{n+k}, \overline{1}\diagdown\underline{n+l} \\ \vdots \qquad\qquad \vdots \qquad\qquad \vdots \qquad\qquad \vdots \\ n\diagdown 1 \ldots\ldots n\diagdown\underline{n-2}, n\diagdown\underline{n+k}, n\diagdown\underline{n+l} \end{vmatrix} \neq 0;$

wherein $k$ and $l$ are any 2 of the numbers 0, 1, ... $r$;

let $x_{n+l}$ have the arbitrary value, 1, assigned to it; and let each of the other Variables $x_n, \ldots x_{n+r}$, except $x_{n+k}$, have the arbitrary value zero assigned to it; and let the Variable $x_{n-1}$ have a possible value assigned to it, and call this value "$a$";

then there are $n$ Equations containing $\overline{n-1}$ Variables, and such that their $B$-Block

$$=\begin{vmatrix} \overline{1}\backslash 1\ldots\ldots\overline{1}\backslash\underline{n-2},\overline{1}\backslash\underline{n+k},\left(\overline{1}\backslash\underline{n-1}.a+\overline{1}\backslash\underline{n+r+1}+\overline{1}\backslash\underline{n+l}\right) \\ \vdots \qquad \vdots \qquad \vdots \qquad \qquad \vdots \\ n\backslash 1\ldots\ldots n\backslash\underline{n-2},n\backslash\underline{n+k},\left(n\backslash\underline{n-1}.a+n\backslash\underline{n+r+1}+n\backslash\underline{n+l}\right) \end{vmatrix};$$

$$=\begin{vmatrix} \overline{1}\backslash 1\ldots\ldots\overline{1}\backslash\underline{n-2},\overline{1}\backslash\underline{n+k},\overline{1}\backslash\underline{n-1} \\ \vdots \qquad \vdots \qquad \vdots \qquad \vdots \\ n\backslash 1\ldots\ldots n\backslash\underline{n-2},n\backslash\underline{n+k},n\backslash\underline{n-1} \end{vmatrix}.a+\begin{vmatrix} \overline{1}\backslash 1\ldots\ldots\overline{1}\backslash\underline{n-2},\overline{1}\backslash\underline{n+k},\overline{1}\backslash\underline{n+r+1} \\ \vdots \qquad \vdots \qquad \vdots \qquad \vdots \\ n\backslash 1\ldots\ldots n\backslash\underline{n-2},n\backslash\underline{n+k},n\backslash\underline{n+r+1} \end{vmatrix}$$

$$+\begin{vmatrix} \overline{1}\backslash 1\ldots\ldots\overline{1}\backslash\underline{n-2},\overline{1}\backslash\underline{n+k},\overline{1}\backslash\underline{n+l} \\ \vdots \qquad \vdots \qquad \vdots \qquad \vdots \\ n\backslash 1\ldots\ldots n\backslash\underline{n-2},n\backslash\underline{n+k},n\backslash\underline{n+l} \end{vmatrix};$$

$$= \qquad 0 \qquad + \qquad 0 \qquad + \qquad \text{do.}$$

∴ it $\neq 0$;

∴ the Equations are inconsistent; (PROP. III.

which is contrary to the hypothesis;

∴ the square Blocks, formed by taking the first $\overline{n-2}$ columns along with any 2 of the columns from the $n^{\text{th}}$ to the $\overline{n+r}|^{\text{th}}$, are all evanescent;

and the same thing may be proved for any $\overline{n-2}$ of the first $\overline{n-1}$ columns;

∴ the square Blocks, formed by taking any $\overline{n-2}$ of the first $\overline{n-1}$ columns along with any 2 of the next $\overline{r+1}$ columns, are all evanescent;

that is, all the principal Minors of the $V$-Block, containing $\overline{n-2}$ of its first $\overline{n-1}$ columns, are evanescent.

The same thing may be proved for $\overline{n-3}$ of these columns, for $\overline{n-4}$ of them, and so on; and finally for all principal Minors, of the $V$-Block, not containing any of its first $\overline{n-1}$ columns.

Therefore $\|V\|=0$.

Therefore $\|B\|=0$. (PROP. XV.

Therefore, if there be, &c. Q. E. D.

## Proposition XIX. Th.

If there be $n$ homogeneous Equations, containing $n$ Variables; and if there be a set of values, for the Variables, which are not all zero (so that at least 2 of them are actual): then $V=0$.

For if not, let $V\neq 0$;
then the values for the Variables are all zero; (Prop. I. Cor.
which is contrary to the hypothesis.

Therefore, if there be, &c. Q. E. D.

## Corollary to Prop. XIX.

If there be $n$ homogeneous Equations, containing $\overline{n-r}$ Variables; and if there be a set of values, for the Variables, which are not all zero (so that at least 2 of them are actual): then $\|V\|=0$.

# CHAPTER IV.

## *TESTS FOR CONSISTENCY OF EQUATIONS.*

### DEFINITIONS.

I.

If there be a condition, or set of conditions, such that, when it is all fulfilled, a certain other condition is also fulfilled: it is said to be **a sufficient test** of that other condition.

II.

And if it be such that, when any part of it is not fulfilled, a certain other condition is not fulfilled: it is said to be **a necessary test** of that other condition.

### CONVENTION.

When a condition, or set of conditions, is said to be a test of a certain other condition, let it be understood that it is sufficient and necessary, unless it be otherwise stated.

### PROPOSITION I. TH.

If there be 2 conditions, whereof the first is a test of the second: the second is likewise a test of the first.

Since the first is a *sufficient* test of the second;

∴ if the first be fulfilled, so is the second;

∴ if the second be not fulfilled, neither is the first;
∴ the second is a *necessary* test of the first.
Again, since the first is a *necessary* test of the second;
∴ if the first be not fulfilled, neither is the second;
∴ if the second be fulfilled, so is the first;
∴ the second is a *sufficient* test of the first.

Therefore, if there be, &c. Q. E. D.

## Proposition II. Th.

If there be given $n$ Equations, not all homogeneous, containing Variables: a test for their being consistent is that either, first, there is one of them such that, when it is taken along with each of the remaining Equations successively, each pair of Equations, so formed, has its $B$-Block evanescent; or, secondly, there are $m$ of them, where $m$ is one of the numbers $2 \ldots\ldots n$, which contain at least $m$ Variables, and have their $V$-Block not evanescent, and are such that, when they are taken along with each of the remaining Equations successively, each set of Equations, so formed, has its $B$-Block evanescent.

Let the test be fulfilled;
in the first case, the Equations are identical; (Ch. III. Prop. VI.
in the second, they are consistent; (Ch. III. Props. I, II, IX, X.
∴ the test is *sufficient*.

Next, let it be not fulfilled;
then there are 2 or more of the Equations, which have their $V$-Block evanescent, but not their $B$-Block;
∴ these Equations are inconsistent; (Ch. III. Props. III, IV, V.
∴ the test is *necessary*.

Therefore, if there be, &c. Q. E. D *.

---

* *Prop. II.* From this Proposition we may deduce a general process for analysing a set of Equations containing Variables. As, however, in the practical application of such a process, it is necessary to test the evanescence of certain Blocks, and as we have at present no convenient method of doing this, the subject is deferred till we come to the Chapter on 'Tests of Evanescence of Blocks.'

### Proposition III. Th.

If there be given 2 Equations containing Variables: a test for their being identical is that $\|B\| = 0$.

Let the test be fulfilled;
then the Equations are identical; (Ch. III. Prop. VI.
∴ it is *sufficient*.

Next, let it be not fulfilled;
in the case where $\|V\| = 0$, the Equations are inconsistent;
(Ch. III. Props. III, IV, V.
in the case where $|V| \neq 0$, they are not identical; (Ch. III. Prop. VII.
∴ the test is *necessary*.

Therefore, if there be, &c. Q. E. D.

### Proposition IV. Th.

If there be given $n$ homogeneous Equations, containing not more than $n$ Variables; a test for there being, for the Variables, a set of values which are not all zero (so that at least 2 of them are actual) is that $\|V\| = 0$.

Let the test be fulfilled;
then there is such a set of values; (Ch. III. Prop. XI.
∴ it is *sufficient*.

Next, let it be not fulfilled;
then there is only one set of values for the Variables; (Ch. III. Prop. I.
and these must each be zero; (Ch. III. Ax. II.
∴ the test is *necessary*.

Therefore, if there be, &c. Q. E. D.

# CHAPTER V.

## *ANALYSIS OF BLOCKS.*

### SECTION I.

*Evanescence of Blocks under given conditions.*

PROPOSITION I. TH.

If, in a square Block, the oblong Block, consisting of 2 or more of its rows or columns, be evanescent: the first Block is also evanescent.

First, let the evanescent oblong Block consist of rows.

Let the rows which constitute it be placed last;

now the Determinant of the first Block may be resolved into terms, each consisting of one of the Elements of the first row, multiplied by the Determinant of one of the principal Minors of the oblong Block formed by erasing the first row; (CH. II. PROP. I. COR. 1.

and each of these Determinants may be in like manner resolved into terms, each containing as a factor the Determinant of one of the principal Minors of the oblong Block formed by erasing the first two rows;

and this process may be repeated, until finally the Determinant of the first Block is resolved into terms, each containing as a factor the Determinant of one of the principal Minors of the evanescent oblong Block;

but each of these vanishes by hypothesis;

∴ the Determinant of the whole Block vanishes.

Similarly, if the evanescent oblong Block consists of columns.

Therefore, if in a square Block, &c. Q. E. D *.

## COROLLARY TO PROP. I.

If, in an oblong Block, the oblong Block, consisting of 2 or more of its longitudinals, be evanescent: the first Block is also evanescent.

## PROPOSITION II. TH.

If there be an oblong Block, having one of its secondary Minors not evanescent; and if, of its principal Minors, each one, which contains that secondary Minor, be evanescent: the whole Block is evanescent.

Call the length of the Block $\overline{n+r}$, and its width $n$.

Let the Block be so placed that its laterals are rows, and let the longitudinals, which contain the non-evanescent Minor, be placed first; and let each of these longitudinals be multiplied throughout by the symbol of a Variable; and let each lateral be equated to zero; and let the $\overline{n+r}$ Equations, so formed, be represented by

---

* *Prop. I.* Thus, if, in the square Block $\left\{\begin{matrix} a & b & c & d \\ e & f & g & h \\ j & k & l & m \\ n & p & q & r \end{matrix}\right\}$, it be given that $\left\| \begin{matrix} j & k & l & m \\ n & p & q & r \end{matrix} \right\| = 0$: the whole Block is evanescent.

$$\text{For } \left\{\begin{matrix} a & b & c & d \\ e & f & g & h \\ j & k & l & m \\ n & p & q & r \end{matrix}\right\} = a . \begin{vmatrix} f & g & h \\ k & l & m \\ p & q & r \end{vmatrix} - b . \begin{vmatrix} e & g & h \\ j & t & m \\ n & q & r \end{vmatrix} + \&c.$$

$$= af . \begin{vmatrix} l & m \\ g & r \end{vmatrix} - ag . \begin{vmatrix} k & m \\ p & r \end{vmatrix} + \&c.$$

$$\overline{1}\backslash 1.x_1 + \ldots\ldots + \overline{1}\backslash \underline{n-1}.x_{n-1} + \overline{1}\backslash n = 0,$$

&c.

$$\overline{n+r}\backslash 1.x_1 + \ldots\ldots + \overline{n+r}\backslash \underline{n-1}.x_{n-1} + \overline{n+r}\backslash n = 0;$$

then there are $\overline{n+r}$ Equations, containing $\overline{n-1}$ Variables; and there are among them $\overline{n-1}$ Equations, which have their *V*-Block not evanescent; and, when such a set of $\overline{n-1}$ Equations is selected and taken along with each of the remaining Equations successively, each set of $n$ Equations, so formed, has its *B*-Block evanescent;

∴ the $\overline{n+r}$ Equations are consistent; (CH. III. PROP. VIII. COR.

∴ their *B*-Block is evanescent. (CHAP. III. PROP. XIII. COR.

Therefore, if there be, &c. Q. E. D*.

## PROPOSITION III. TH.

If there be a Block, having one of its Minors of the $k^{th}$ degree, where $k$ is less than the degree of a secondary Minor, not evanescent; and if, of the oblong Blocks formed from it by selecting $\overline{k+1}$ of its longitudinals, each one, which contains that non-evanescent Minor, be evanescent: every other oblong Block, so formed, is evanescent. And the same is true of its laterals.

Call the two dimensions of the Block '*m*' and '*n*'.

Let the Block be placed in either position, and let the $k$ rows, and also the $k$ columns, which contain the non-evanescent Minor, be placed first; and let each of the columns, except the last, be multiplied throughout by the symbol of a Variable; and let each row be equated to zero; and let the Equations, so formed, be represented by

---

* *Prop. II.* Thus, if, in the oblong Block $\begin{Bmatrix} a & b & c & d \\ e & f & g & h \\ j & k & l & m \end{Bmatrix}$, it be given that $\begin{vmatrix} b & d \\ k & m \end{vmatrix} \neq 0$, and that $\begin{vmatrix} a & b & d \\ e & f & h \\ j & k & m \end{vmatrix} = 0$, and $\begin{vmatrix} b & c & d \\ f & g & h \\ k & l & m \end{vmatrix} = 0$: the whole Block is evanescent.

$$\overline{1}\backslash 1.x_1+\ldots\ldots+\overline{1}\backslash k.x_k+\ldots\ldots+\overline{1}\backslash\underline{m-1}.x_{m-1}+\overline{1}\backslash m = 0,$$

&c.

$$\overline{k}\backslash 1.x_1+\ldots\ldots+\overline{k}\backslash k.x_k+\ldots\ldots+\overline{k}\backslash\underline{m-1}.x_{m-1}+\overline{k}\backslash m = 0,$$

$$\overline{k+1}\backslash 1.x_1+\ldots\ldots+\overline{k+1}\backslash k.x_k+\ldots\ldots+\overline{k+1}\backslash\underline{m-1}.x_{m-1}+\overline{k+1}\backslash m = 0,$$

&c.

$$\overline{n}\backslash 1.x_1+\ldots\ldots+\overline{n}\backslash k.x_k+\ldots\ldots+\overline{n}\backslash\underline{m-1}.x_{m-1}+\overline{n}\backslash m = 0;$$

where $m$ is $>$, $=$, or $< n$; and where $\begin{vmatrix} \overline{1}\backslash 1 \ldots\ldots \overline{1}\backslash k \\ \vdots \qquad \vdots \\ \overline{k}\backslash 1 \ldots\ldots \overline{k}\backslash k \end{vmatrix} \neq 0$;

then the first $\overline{k+1}$ Equations contain $\overline{m-1}$ Variables, and there are among them $k$ Equations whose *V*-Block is not evanescent; and the *B*-Block of these $\overline{k+1}$ Equations is evanescent;

also, since $\overline{m-1} > k$, $\quad\therefore\quad \overline{m-1} \not< \overline{k+1}$.

First, let $\overline{m-1} = \overline{k+1}$;

then the Equations are consistent; and the Variable $x_{m-1}$ may have an arbitrary value assigned to it; and the $\overline{k+1}|^{\text{th}}$ Equation is dependent on the first $k$ Equations; (CH. III. PROP. IX.

also, if any of the remaining Equations be substituted for the $\overline{k+1}|^{\text{th}}$ Equation, the same thing may be proved;

$\therefore$ the $n$ Equations are consistent, and the Variable $x_{m-1}$ may have an arbitrary value assigned to it;

$\therefore$ if any set of $\overline{k+1}$ Equations be selected, the same thing is true of them;

$\therefore$ any such set has its *B*-Block evanescent. (CH. III. PROP. XVII.

Secondly, let $\overline{m-1} = \overline{k+1}+r$;

then the Equations are consistent; and the $\overline{r+1}$ Variables, $x_{k+1}, \ldots x_{m-1}$, may have arbitrary values assigned to them; and the $\overline{k+1}|^{\text{th}}$ Equation is dependent on the first $k$ Equations; (CH. III. PROP. IX. COR.

also, if any of the remaining Equations be substituted for the $\overline{k+1}|^{\text{th}}$ Equation, the same thing may be proved;

∴ the $n$ Equations are consistent, and these $\overline{r+1}$ Variables may have arbitrary values assigned to them;

∴ if any set of $\overline{k+1}$ Equations be selected, the same thing is true of them;

∴ any such set has its $B$-Block evanescent. (CH. III. PROP. XVIII.

Therefore, if there be, &c. Q. E. D*.

## COROLLARY TO PROP. III.

If there be a Block, having one of its Minors of the $k^{th}$ degree not evanescent, where $k$ is less than the degree of a secondary Minor; and if, of its Minors of the $\overline{k+1}|^{th}$ degree, each one, which contains that non-evanescent Minor, be evanescent: every other Minor of that degree is evanescent.

## PROPOSITION IV. TH.

If there be a Block containing 2 rows and 2 or more columns; and if, in every column, the 1st term bear to the 2nd a constant ratio: the Block is evanescent.

Let the Block be represented by

$$\begin{Bmatrix} a_1, a_2, \ldots\ldots a_n \\ b_1, b_2, \ldots\ldots b_n \end{Bmatrix};$$

and let it be given that

$$a_1 : b_1 :: a_2 : b_2 :: \&c. :: a_n : b_n;$$
$$:: k : 1 \text{ (say)};$$

$$\therefore \quad a_1 = kb_1, \quad a_2 = kb_2, \&c., \quad a_n = kb_n;$$

---

* *Prop. III.* Thus, if, in the Block $\begin{Bmatrix} a & b & c & d & e \\ f & g & h & j & k \\ l & m & n & p & q \\ r & s & t & u & v \end{Bmatrix}$, it be given that $\begin{vmatrix} h & k \\ n & q \end{vmatrix} \neq 0$, and that $\begin{Vmatrix} a & b & c & d & e \\ f & g & h & j & k \\ l & m & n & p & q \end{Vmatrix} = 0$, and $\begin{Vmatrix} f & g & h & j & k \\ l & m & n & p & q \\ r & s & t & u & v \end{Vmatrix} = 0$: then $\begin{Vmatrix} a & b & c & d & e \\ f & g & h & j & k \\ r & s & t & u & v \end{Vmatrix} = 0$, and $\begin{Vmatrix} a & b & c & d & e \\ l & m & n & p & q \\ r & s & t & u & v \end{Vmatrix} = 0$.

now $\begin{Vmatrix} kb_1, & kb_2, \ldots\ldots & kb_n \\ b_1, & b_2, \ldots\ldots & b_n \end{Vmatrix} = 0$, since every principal Minor of this Block, if its first row be divided by $k$, has 2 rows identical;

that is, $\begin{Vmatrix} a_1, & a_2, \ldots\ldots & a_n \\ b_1, & b_2, \ldots\ldots & b_n \end{Vmatrix} = 0.$

Therefore, if there be, &c. Q. E. D.

## PROPOSITION V. TH.

If there be a Block containing 3 rows and 3 or more columns; and if, in every column, the difference between the 1st and 2nd terms bears to the 3rd term a constant ratio: the Block is evanescent.

Let the Block be represented by $\begin{Bmatrix} a_1, & a_2, \ldots\ldots & a_n \\ b_1, & b_2, \ldots\ldots & b_n \\ c_1, & c_2, \ldots\ldots & c_n \end{Bmatrix}$; and let it be given that

$$(a_1-b_1) : c_1 :: (a_2-b_2) : c_2 :: \&c. :: (a_n-b_n) : c_n;$$
$$:: (k : 1 \text{ (say)};$$

$$\therefore\ (a_1-b_1) = kc_1,\ (a_2-b_2) = kc_2,\ \&c.,\ (a_n-b_n) = kc_n;$$

now $\begin{Vmatrix} kc_1, & kc_2, \ldots\ldots & kc_n \\ b_1, & b_2, \ldots\ldots & b_n \\ c_1, & c_2, \ldots\ldots & c_n \end{Vmatrix} = 0$, since every principal Minor of this Block, if its 1st row be divided by $k$, has 2 rows identical;

that is, $\begin{Vmatrix} (a_1-b_1), & (a_2-b_2), \ldots\ldots & (a_n-b_n) \\ b_1, & b_2, \ldots\ldots & b_n \\ c_1, & c_2, \ldots\ldots & c_n \end{Vmatrix} = 0;$

$\therefore$, adding to the terms of the 1st row those of the 2nd,

$$\begin{Vmatrix} a_1, & a_2, \ldots\ldots & a_n \\ b_1, & b_2, \ldots\ldots & b_n \\ c_1, & c_2, \ldots\ldots & c_n \end{Vmatrix} = 0. \qquad \text{(CH. II. PROP. III. COR. 3.}$$

Therefore, if there be, &c. Q. E. D.

## PROPOSITION VI. TH.

If there be a Block containing 3 rows and 3 or more columns; and if, in every column, the difference between the 1st and 2nd terms bears to the difference between the 2nd and 3rd a constant ratio: the Block is evanescent.

Let the Block be represented by $\begin{Bmatrix} a_1, & a_2, & \ldots\ldots & a_n \\ b_1, & b_2, & \ldots\ldots & b_n \\ c_1, & c_2, & \ldots\ldots & c_n \end{Bmatrix}$; and let it be given that

$$(a_1-b_1):(b_1-c_1)::(a_2-b_2):(b_2-c_2):: \&c. :: (a_n-b_n):(b_n-c_n);$$
$$::k:1 \text{ (say)};$$

$$\therefore \quad (a_1-b_1) = k(b_1-c_1), \quad (a_2-b_2) = k(b_2-c_2), \ \&c., \quad (a_n-b_n) = k(b_n-c_n);$$

now $\begin{Vmatrix} k(b_1-c_1), & k(b_2-c_2), & \ldots\ldots & k(b_n-c_n) \\ (b_1-c_1), & (b_2-c_2), & \ldots\ldots & (b_n-c_n) \\ c_1, & c_2, & \ldots\ldots & c_n \end{Vmatrix} = 0$, since every principal Minor of this Block, if its 1st row be divided by $k$, has 2 rows identical;

that is, $\begin{Vmatrix} (a_1-b_1), & (a_2-b_2), & \ldots\ldots & (a_n-b_n) \\ (b_1-c_1), & (b_2-c_2), & \ldots\ldots & (b_n-c_n) \\ c_1, & c_2, & \ldots\ldots & c_n) \end{Vmatrix} = 0;$

$\therefore$, adding to the terms of the 1st row those of the 2nd and 3rd, and to the terms of the 2nd row those of the 3rd,

$$\begin{Vmatrix} a_1, & a_2, & \ldots\ldots & a_n \\ b_1, & b_2, & \ldots\ldots & b_n \\ c_1, & c_2, & \ldots\ldots & c_n \end{Vmatrix} = 0. \qquad \text{(CH. II. PROP. III. COR. 3.}$$

Therefore, if there be, &c. Q. E. D.

## CHAPTER V. *(Continued.)*

### SECTION II.

*Properties of Blocks under given conditions of evanescence.*

#### PROPOSITION VII. TH.

If there be a Block containing 2 rows and 2 or more columns; and if it be evanescent: then, in every column, the first term bears to the second a constant ratio.

Let the Block be represented by $\left\{ \begin{matrix} a_1, a_2, \ldots\ldots a_n \\ b_1, b_2, \ldots\ldots b_n \end{matrix} \right\}$;

$\because \begin{vmatrix} a_1, a_2 \\ b_1, b_2 \end{vmatrix} = 0$;

$\therefore a_1 b_2 = a_2 b_1$;

$\therefore \dfrac{a_1}{b_1} = \dfrac{a_2}{b_2}$;

$\therefore a_1 : b_1 :: a_2 : b_2 ::$ (by symmetry) &c. $:: a_n : b_n$.

Therefore, if there be, &c. Q. E. D.

#### PROPOSITION VIII. TH.

If there be a Block containing 3 rows and 3 or more columns; and if, in one of its columns, the 1st and 2nd terms be equal and the 3rd zero; and if it be evanescent: then, in every column, the difference between the 1st and 2nd terms bears to the 3rd term a constant ratio.

Let the Block be represented by $\left\{\begin{matrix} a_1, & a_2, \ldots\ldots & h \\ b_1, & b_2, \ldots\ldots & h \\ c_1, & c_2, \ldots\ldots & 0 \end{matrix}\right\}$;

$$\because \quad \begin{vmatrix} a_1, & a_2, & h \\ b_1, & b_2, & h \\ c_1, & c_2, & 0 \end{vmatrix} = 0;$$

$\therefore$, subtracting from the terms of the 1st row those of the 2nd,

$$\begin{vmatrix} (a_1-b_1), & (a_2-b_2), & 0 \\ b_1, & b_2, & h \\ c_1, & c_2, & 0 \end{vmatrix} = 0; \qquad \text{(CH. II. PROP. III. COR. 3.}$$

$$\therefore \quad \begin{vmatrix} (a_1-b_1), & (a_2-b_2) \\ c_1, & c_2 \end{vmatrix} = 0; \qquad \text{(CH. II. PROP. I. COR. 2.}$$

$$\therefore \quad (a_1-b_1) : c_1 :: (a_2-b_2) : c_2; \qquad \text{(PROP. VII.}$$

$$:: (a_3-b_3) : c_3; \quad \text{(by symmetry)}$$

$$:: \&c.$$

Therefore, if there be, &c. Q. E. D.

## PROPOSITION IX. TH.

If there be a Block containing 3 rows and 3 or more columns; and if, in one of its columns, the 3 terms be equal; and if the Block be evanescent: then, in every column, the difference between the 1st and 2nd terms bears to the difference between the 2nd and 3rd a constant ratio.

Let the Block be represented by $\left\{\begin{matrix} a_1, & a_2, \ldots\ldots & h \\ b_1, & b_2, \ldots\ldots & h \\ c_1, & c_2, \ldots\ldots & h \end{matrix}\right\}$;

$$\because \quad \begin{vmatrix} a_1, & a_2, & h \\ b_1, & b_2, & h \\ c_1, & c_2, & h \end{vmatrix} = 0;$$

$\therefore$, subtracting from the terms of the 1st row those of the 2nd, and from the terms of the 2nd those of the third,

$$\begin{vmatrix} (a_1-b_1), & (a_2-b_2), & 0 \\ (b_1-c_1), & (b_2-c_2), & 0 \\ c_1, & c_2, & h \end{vmatrix} = 0; \qquad \text{(Ch. II. Prop. III. Cor. 3.}$$

$$\therefore \quad \begin{vmatrix} (a_1-b_1), & (a_2-b_2) \\ (b_1-c_1), & (b_2-c_2) \end{vmatrix} = 0; \qquad \text{(Ch. II. Prop. I. Cor. 2.}$$

$$\therefore \quad (a_1-b_1) : (b_1-c_1) :: (a_2-b_2) : (b_2-c_2);$$
$$:: (a_3-b_3) : (b_3-c_3); \text{ (by symmetry)}$$
$$:: \&c.$$

Therefore, if there be, &c. Q. E. D.

## Proposition X. Th.

If there be an oblong Block, whose length exceeds its breadth by unity; and if one of its principal Minors be non-evanescent: at least one other is also non-evanescent.

In the lateral, which is not included in the non-evanescent principal Minor, let an actual term be selected; and let the Elements of the longitudinal, which contains the selected term, be each multiplied by the Determinant of the Minor formed by erasing the lateral containing that Element;

then the sum of these products, affected with + and − alternately, is zero; (Ch. II. Prop. III. Cor. 2.

$\therefore$ if one of them be actual, at least one other is actual;

$\therefore$ at least one other principal Minor of the oblong Block is non-evanescent.

Therefore, if there be, &c. Q. E. D.

# CHAPTER VI.

## *TESTS FOR EVANESCENCE OF BLOCKS.*

### PROPOSITION I. TH.

If there be given an oblong Block, having one of its secondary Minors not evanescent; a test of its being evanescent is that, of its principal Minors, each one, which contains that secondary Minor, is evanescent.

Let the test be fulfilled;
then the given Block is evanescent; (CH. V. PROP. I.
∴ it is *sufficient.*

Next, let it be not fulfilled;
then the given Block is not evanescent;
∴ it is *necessary.*

Therefore, if there be, &c. Q. E. D *.

### PROPOSITION II. TH.

If there be given a Block: a test of its being evanescent is that either every Element of it is zero; or there are 2 or more

---

* *Prop. I.* Thus, in the oblong Block $\left\{\begin{matrix} 2 & 1 & 3 & -1 & 0 \\ -1 & 3 & -5 & 4 & -7 \\ 1 & 0 & 2 & -1 & 1 \end{matrix}\right\}$, we have $\begin{vmatrix} 2 & 1 \\ -1 & 3 \end{vmatrix} \neq 0$,
$\begin{vmatrix} 2 & 1 & 3 \\ 1 & 3 & -5 \\ 1 & 0 & 2 \end{vmatrix} = 0$, $\begin{vmatrix} 2 & 1 & -1 \\ -1 & 3 & 4 \\ 1 & 0 & -1 \end{vmatrix} = 0$, $\begin{vmatrix} 2 & 1 & 0 \\ -1 & 3 & -7 \\ 1 & 0 & 1 \end{vmatrix} = 0$. Hence the whole Block is evanescent.

of its longitudinals, which form a Block, having one of its secondary Minors not evanescent, and such that, of its principal Minors, each one, which contains that secondary Minor, is evanescent.

Let the test be fulfilled ;

in the case where each Element of the Block is zero, it is plain that the Block is evanescent ;

in the case where 2 or more of its longitudinals, but not all of them, form such a Block, the Block, so formed, is evanescent ; (CH. V. PROP. II.

∴ the given Block is evanescent ; (CH. V. PROP. I.

in the case where all the longitudinals of the given Block form such a Block, it is evanescent ;

∴ the test is *sufficient*. (CH. V. PROP. II.

Next, let it be not fulfilled ;

then one of the Elements of the Block is actual ;

∴ any Block, formed of 2 of the longitudinals of the given Block, has one of its principal Minors not evanescent ;

for otherwise, it must have all of its principal Minors evanescent, and, since it necessarily has one of its secondary Minors not evanescent, the test would be fulfilled ;

∴ every Block, formed of 2 of the longitudinals of the given Block, is not evanescent ;

then it may be proved, in the same manner, that every Block formed of 3 of the longitudinals of the given Block, is not evanescent, and so on, up to the given Block itself ;

∴ the given Block is not evanescent ;

∴ the test is *necessary*.

Therefore, if there be, &c. Q. E. D *.

---

* *Prop. II.* Hence the oblong Block $\left\{\begin{matrix} 2 & 1 & 3 & -1 & 0 \\ -1 & 3 & -5 & 4 & -7 \\ 1 & 0 & 2 & -1 & 1 \\ 3 & 4 & -2 & 5 & -1 \end{matrix}\right\}$ is evanescent, since its first 3 longitudinals form the Block discussed in the last Note.

We are now in a position to describe, and apply, a general process for analysing a set of Equations containing Variables. This will be found in Appendix I.

## PROPOSITION III. TH.

If there be given a Block: a test for the evanescence of every oblong Block, formed from it by selecting $h$ of its laterals, is that either every Element of it is zero, or that it has a non-evanescent Minor of the $k^{th}$ degree, where $k$ is less than $h$, such that, of the oblong Blocks formed from it by selecting $\overline{k+1}$ of its laterals, each one, which contains that non-evanescent Minor, is evanescent. And the test for the longitudinals of the given Block is similar to this.

Let the test be fulfilled;

in the case where every Element of the given Block is zero, it is evident that every oblong Block, formed from it by selecting $h$ of its laterals, is evanescent;

in the other case, the same results follow; (CH. V. PROP. III.

∴ the test is *sufficient.*

Next, let it be not all fulfilled;

then the given Block contains one or more actual Elements, and each one of its non-evanescent Minors of the $k^{th}$ degree, where $k$ is less than $h$, is such that, among the oblong Blocks, formed from the given Block by selecting $\overline{k+1}$ of its laterals, there is one, containing that non-evanescent Minor, which is itself non-evanescent;

∴ this is true when $k = 1$;

∴ among the oblong Blocks, formed from the given Block by selecting 2 of its laterals, there is one non-evanescent;

∴ the given Block has a non-evanescent Minor of the $2^{nd}$ degree;

∴ among the oblong Block, formed from the given Blocks by selecting 3 of its laterals, there is one non-evanescent;

and the same thing may be proved for all values of $k$ up to $\overline{h-1}$;

∴ among the oblong Blocks, formed from the given Block by selecting $h$ of its laterals, there is one non-evanescent.

∴ the test is *necessary.*

Therefore this is proved to be a test for the laterals of the given Block;

and a similar set of conditions may be proved, in like manner, to be a test for its longitudinals.

Therefore, if there be, &c. Q. E. D.

# CHAPTER VII.

## *GEOMETRICAL ANALYSIS.*

### Convention I.

When mention is made of a Point, a Line, or a Plane, let it be understood that the words "at a finite distance" are to be added, unless it be otherwise expressed.

## SECTION I.

### *Plane Geometry.*

### Definition I.

In the Trilinear System, the Equation

$$a\alpha + b\beta + c\gamma - 2M = 0$$

is called the **systematic** Equation.

### Convention II.

In the Trilinear System, when the coordinates of a Point are given, let it be understood that they satisfy the Systematic Equation.

### Proposition I. Th.

If there be given an Equation of the first degree;
first, in the Cartesian System, viz. —

$$Ax + By + C = 0;$$

and

(1) If either $A$, or $B$, $\neq 0$;
then the Equation represents one real Line, and one only.

(2) If $A = B = 0$, but $C \neq 0$;
then it does not represent a real Line.

(3) If $A = B = C = 0$;
then it represents the Plane of reference:

secondly, in the Trilinear System, viz.—

$$A\alpha + B\beta + C\gamma + D = 0,$$

the systematic Equation being $a\alpha + b\beta + c\gamma - 2M = 0$;

and

(1) If $|V| \neq 0$;
then the 2 Equations are consistent; (Ch. III. Prop. II.
∴ the given Equation represents a real Line.
also there is one Variable to which an arbitrary value may be given, and, for each such arbitrary value, there is only one value for each of the other Variables; (Ch. III. Prop. II.
∴ the given Equation represents one real Line, and one only.

(2) If $\|V\| = 0$, but $|B| \neq 0$;
then the 2 Equations are inconsistent; (Ch. III. Prop. V.
∴ the given Equation does not represent a real Line.

(3) If $\|B\| = 0$; (whence also $\|V\| = 0$);
then the 2 Equations are identical; (Ch. III. Prop. VI.
∴ the given Equation represents the Plane of reference.

CONVENTIONS (*continued*).

III.

When an Equation to a Line is given in the form

$$Ax + By + C = 0,$$

let it be understood that either $A$, or $B$, $\neq 0$ ; when in the form

$$A\alpha + B\beta + C\gamma + D = 0,$$

the systematic Equation being $a\alpha + b\beta + c\gamma - 2M = 0$ ;

let it be understood that $|V| \neq 0$.

IV.

In the Trilinear System, when mention is made of the $V$-Block, or $B$-Block, of any number of Equations to Lines : let it be understood that, in forming such Block, the systematic Equation is always taken along with them.

V.

When 2 Lines are said to **intersect in a Point at an infinite distance,** let it be understood that they are parallel.

---

PROPOSITION II. TH.

If the Equation to a line, passing through the origin or a vertex of reference, be given in the form

$$\frac{x}{l} = \frac{y}{m} = 0; \qquad \text{or} \qquad \frac{\alpha}{l} = \frac{\beta}{m}:$$

this may be written in the form

$$\begin{vmatrix} x, & y \\ l, & m \end{vmatrix} = 0, \qquad \text{or} \qquad \begin{vmatrix} \alpha, & \beta \\ l, & m \end{vmatrix} = 0, \qquad \text{(CH. V. PROP. IV.}$$

Again, if the Equation to a Line, passing through a given Point, be given in the form

$$\frac{x-x'}{l} = \frac{y-y'}{m}, \quad \text{or} \quad \frac{\alpha-\alpha'}{l} = \frac{\beta-\beta'}{m}:$$

this may be written in the form

$$\begin{vmatrix} x, & y, & 1 \\ x', & y', & 1 \\ l, & m, & 0 \end{vmatrix} = 0, \quad \text{or} \quad \begin{vmatrix} \alpha, & \beta, & 1 \\ \alpha', & \beta', & 1 \\ l, & m, & 0 \end{vmatrix} = 0. \quad \text{(Ch. V. Prop. V.}$$

Again, if the Equation to a Line, passing through 2 given Points, be given in the form

$$\frac{x-x_1}{x_1-x_2} = \frac{y-y_1}{y_1-y_2}, \quad \text{or} \quad \frac{\alpha-\alpha_1}{\alpha_1-\alpha_2} = \frac{\beta-\beta_1}{\beta_1-\beta_2}:$$

this may be written in the form

$$\begin{vmatrix} x, & y, & 1 \\ x_1, & y_1, & 1 \\ x_2, & y_2, & 1 \end{vmatrix} = 0, \quad \text{or} \quad \begin{vmatrix} \alpha, & \beta, & 1 \\ \alpha_1, & \beta_1, & 1 \\ \alpha_2, & \beta_2, & 1 \end{vmatrix} = 0, \quad \text{(Ch. V. Prop. VI.}$$

## Proposition III. Th.

If there be given 2 Lines, represented, in the Cartesian System, by

$$A_1x + B_1y + C_1 = 0,$$
$$A_2x + B_2y + C_2 = 0;$$

or, in the Trilinear system, by

$$A_1\alpha + B_1\beta + C_1\gamma + D_1 = 0,$$
$$A_2\alpha + B_2\beta + C_2\gamma + D_2 = 0,$$

the systematic Equation being $a\alpha + b\beta + c\gamma - 2M = 0$;

and

(1) If $V \neq 0$;

then the Equations are consistent, and there is only one set of values for the Variables; (Ch. III. Prop. I.

∴ the 2 Lines intersect in one Point, and in one only.

(2) If $V=0$, but $|B| \neq 0$;

then the Equations are inconsistent; (CH III. PROP. IV.

$\therefore$ the 2 Lines are parallel.

(3) If $\|B\|=0$; (whence also $V=0$);

then, in the Cartesian System, the Equations are identical;

(CH. III. PROP. VI.

also, in the Trilinear System, since either of the Equations to the Lines, taken with the systematic Equation, have their $V$-Block not evanescent;

(CONV. II.

$\therefore$ the 3 Equations are consistent, and either of the Equations to the Lines is dependent on the other Equations; (CH. III. PROP. VIII.

$\therefore$ whatsoever Point lies on one of the 2 Lines, lies also on the other; (CH. III. DEF. VI.

$\therefore$ in either System, the 2 Lines coincide*.

From (2) and (3) may be deduced

(4) If $V=0$; the 2 Lines have the same direction.

## PROPOSITION IV. TH.

If there be given 3 Lines, represented, in the Cartesian System, by

$$A_1x+B_1y+C_1 = 0,$$
$$A_2x+B_2y+C_2 = 0,$$
$$A_3x+B_3y+C_3 = 0;$$

or, in the Trilinear System, by

$$A_1\alpha+B_1\beta+C_1\gamma+D_1 = 0,$$
$$A_2\alpha+B_2\beta+C_2\gamma+D_2 = 0,$$
$$A_3\alpha+B_3\beta+C_3\gamma+D_3 = 0,$$

the systematic Equation being $a\alpha+b\beta+c\gamma-2M=0$;

and

---

* *Prop. III.* (3.) Thus, if the given Lines be

$$\alpha+\beta\quad+1=0,$$
$$\alpha+2\beta+5\gamma-14=0;$$

the systematic Equation being $3\alpha+4\beta+5\gamma-12=0$;

$\|B\|=0$, and therefore the 2 Lines coincide.

(1) If $B \neq 0$; (whence also $|V| \neq 0$, and every 2 of the Equations to the 3 Lines have their $B$-Block not evanescent);

then the Equations are inconsistent; (Ch. III. Prop. III.

$\therefore$ the 3 Lines do not intersect in one Point, but there are 2 of them which intersect in one Point, and in one only.

(Prop. III. (1)

(2) If $B=0$; and if there be, among the Equations to the 3 Lines, 2 which have their $V$-Block not evanescent;

then the Equations are consistent, and there is only one set of values for the Variables; (Ch. III. Prop. VIII.

$\therefore$ the 3 Lines intersect in one Point, and in one only *.

(3) If $B=0$; and if every 2 of the Equations to the 3 Lines have their $V$-Block evanescent; and if there be among them 2 which have their $B$-Block not evanescent;

then 2 of the Lines are parallel; (Prop. III. (2)

and the 3 have the same direction. (Prop. III. (4)

(4) If every 2 of the Equations to the 3 Lines have their $B$-Block evanescent; (whence also $B=0$);

then the 3 Lines coincide. (Prop. III. (3)

From (2), (3), and (4) may be deduced

(5) If $B=0$;

then the 3 Lines intersect in one Point at a finite or infinite distance.

From (2) and (4) may be deduced

---

* *Prop. IV.* (2.) Thus, if the given Lines be

$$\alpha-\beta+\gamma+1=0,$$
$$3\alpha+\beta-2\gamma-4=0,$$
$$2\alpha+9\beta-\gamma-20=0,$$

the systematic Equation being $3\alpha+4\beta+5\gamma-12=0$;

since $B=0$, and $\begin{vmatrix} 1 & -1, & 1 \\ 3 & 1, & -2 \\ 3 & 4, & 1 \end{vmatrix} \neq 0$, the 3 Lines intersect in one Point, and in one only.

(6) If $B=0$; and if either there be, among the Equations to the 3 Lines, 2 which have their $V$-Block not evanescent, or if every 2 have their $B$-Block evanescent;
then the 3 Lines intersect in one Point.

## PROPOSITION V. TH.

If there be given 2 Points, represented, in the Trilinear System, by $(\alpha_1, \beta_1, \gamma_1)$, $(\alpha_2, \beta_2, \gamma_2)$; and if

$$\left\| \begin{matrix} \alpha_1, \beta_1, \gamma_1 \\ \alpha_2, \beta_2, \gamma_2 \end{matrix} \right\| = 0;$$

then $$\left\| \begin{matrix} \alpha_1, \beta_1, \gamma_1, 1 \\ \alpha_2, \beta_2, \gamma_2, 1 \end{matrix} \right\| = 0.$$

For $a\alpha_1 + b\beta_1 + c\gamma_1 - 2M = 0$,

$a\alpha_2 + b\beta_2 + c\gamma_2 - 2M = 0$;

in these Equations, let $a, b, c, -2M$, be considered as Variables;

then there are 2 homogeneous Equations containing 4 Variables; and their $V$-Block is $\left\{ \begin{matrix} \alpha_1, \beta_1, \gamma_1, 1 \\ \alpha_2, \beta_2, \gamma_2, 1 \end{matrix} \right\}$; and there is a set of values, for the Variables, of which the last is actual; and, if the last column of their $V$-Block be omitted, the remaining Block is evanescent;

$\therefore$ the whole Block is evanescent. (CH. III. PROP. XV. COR.

Therefore, if there be given, &c. Q. E. D.

### COROLLARIES TO PROP. V.

1.

Hence the 2 Points are coincident.

For, since $\left| \begin{matrix} \alpha_1, 1 \\ \alpha_2, 1 \end{matrix} \right| = 0$, $\alpha_1 = \alpha_2$; and so of the others.

2.

By a similar process (using Ch. III. Prop. XIV. Cor.), it may be proved that, if there be 3 Points, thus represented, and if

$$\begin{vmatrix} \alpha_1, & \beta_1, & \gamma_1 \\ \alpha_2, & \beta_2, & \gamma_2 \\ \alpha_3, & \beta_3, & \gamma_3 \end{vmatrix} = 0:$$

$$\text{then} \quad \left\Vert \begin{matrix} \alpha_1, & \beta_1, & \gamma_1, & 1 \\ \alpha_2, & \beta_2, & \gamma_2, & 1 \\ \alpha_3, & \beta_3, & \gamma_3, & 1 \end{matrix} \right\Vert = 0.$$

## Proposition VI. Th.

If there be given 3 Points, represented, in the Cartesian System, by $(x_1, y_1)$ &c.; or, in the Trilinear System, by $(\alpha_1, \beta_1, \gamma_1)$, &c.; and if, in each System, the general Equation to a Line be taken, involving 3 undetermined quantities $A$, $B$, $C$; and if the coordinates of the given Points be successively substituted in it; and if $A$, $B$, and $C$ be considered as Variables in the Equations so formed, viz.—

in the Cartesian System,

$$\left.\begin{matrix} Ax_1 + By_1 + C = 0, \\ Ax_2 + By_2 + C = 0, \\ Ax_3 + By_3 + C = 0, \end{matrix}\right\}, \text{ whose } V\text{-Block is } \left\{\begin{matrix} x_1, & y_1, & 1 \\ x_2, & y_2, & 1 \\ x_3, & y_3, & 1 \end{matrix}\right\};$$

or, in the Trilinear System,

$$\left.\begin{matrix} A\alpha_1 + B\beta_1 + C\gamma_1 = 0, \\ A\alpha_2 + B\beta_2 + C\gamma_2 = 0, \\ A\alpha_3 + B\beta_3 + C\gamma_3 = 0, \end{matrix}\right\}, \text{ whose } V\text{-Block is } \left\{\begin{matrix} \alpha_1, & \beta_1, & \gamma_1 \\ \alpha_2, & \beta_2, & \gamma_2 \\ \alpha_3, & \beta_3, & \gamma_3 \end{matrix}\right\};$$

and

(1) If $V \neq 0$;

then the only values for $A$, $B$, $C$ are zero; (Ch. III. Prop. I. Cor.

$\therefore$ the 3 Points do not lie on one Line.

(2) If $V=0$; and if there be, among the 3 Equations, 2 which have their $V$-Block not evanescent;

then there are 2 of the Points which do not coincide, that is, which lie on one Line, and on one only;

also the 3rd Equation is dependent on the others; (CH. III. PROP. VIII.

that is, those values of $A$, $B$, and $C$, which belong to a Line passing through these 2 Points, the same belong also to a Line passing through the 3rd Point;

$\therefore$ the 3 Points lie on one Line, and on one only *.

(3) If every 2 of the 3 Equations have their $V$-Block evanescent; (whence also $V=0$);

then the 3 Points coincide. (PROP. V. COR. I.

From (2) and (3) may be deduced

(4) If $V=0$;

then the 3 Points lie on one Line.

---

* *Prop. VI.* (2.) Thus, in the Cartesian system, if the given Points be (2, 5), (3, −1), (1, 11); since $\begin{vmatrix} 2 & 5 & 1 \\ 3 & -1 & 1 \\ 1 & 11 & 1 \end{vmatrix} = 0$, and $\begin{vmatrix} 2 & 5 & 1 \\ 3 & -1 & 1 \end{vmatrix} \neq 0$, the three Points lie on one Line, and on one only.

Again, in the Trilinear system, if the given Points be (1, 3, −2), (2, −1, 5), (4, 5, 1); since $\begin{vmatrix} 1 & 3 & -2 \\ 2 & -1 & 5 \\ 4 & 5 & 1 \end{vmatrix} = 0$, and $\begin{vmatrix} 1 & 3 & -2 \\ 2 & -1 & 5 \end{vmatrix} \neq 0$, the same result follows.

# CHAPTER VII. *(Continued.)*

## SECTION II.

---

### *Solid Geometry.*

---

DEFINITION II.

In the Quadriplanar System, the Equation

$$a\alpha + b\beta + c\gamma + d\delta - 3M = 0$$

is called the **systematic** Equation.

---

CONVENTION VI.

In the Quadriplanar System, when the coordinates of a Point are given, let it be understood that they satisfy the Systematic Equation.

---

PROPOSITION VII. TH.

If there be given an Equation of the first degree;

first, in the Cartesian System, viz.—

$$Ax + By + Cz + D = 0;$$

and

(1) If either $A$, or $B$, or $C$, and $\neq 0$;

then the Equation represents one real Plane, and one only.

(2) If $A = B = C = 0$, but $D \neq 0$;

then it does not represent a real Plane.

(3) If $A = B = C = D = 0$;

then it represents all Space:

secondly, in the Quadriplanar System, viz.—

$$A\alpha + B\beta + C\gamma + D\delta + E = 0,$$

the systematic Equation being $a\alpha + b\beta + c\gamma + d\delta - 3M = 0$;

and

(1) If $|V| \neq 0$;

then the 2 Equations are consistent; (CH. III. PROP. II.

$\therefore$ the given Equation represents a real Plane;

also there are 2 Variables to which arbitrary values may be given, and, for each such set of arbitrary values, there is only one value for each of the other Variables; (CH. III. PROP. II.

$\therefore$ the given Equation represents one real Plane, and one only.

(2) If $\|V\| = 0$, but $|B| \neq 0$;

then the 2 Equations are inconsistent; (CH. III. PROP. V.

$\therefore$ the given Equation does not represent a real Plane.

(3) If $\|B\| = 0$; (whence also $\|V\| = 0$);

then the 2 Equations are identical; (CH. III. PROP. VI.

$\therefore$ the given Equation represents all Space.

---

## CONVENTIONS (*continued*).

### VII.

When an Equation to a Plane is given in the form

$$Ax + By + Cz + D = 0,$$

let it be understood that either $A$, or $B$, or $C$, $\neq 0$; when in the form

$$A\alpha + B\beta + C\gamma + D\delta + E = 0,$$

the systematic Equation being $a\alpha + b\beta + C\gamma + d\delta - 3M = 0$,

let it be understood that $|V| \neq 0$.

## VIII.

In the Quadriplanar System, when mention is made of the $V$-Block, or $B$-Block, of any number of Equations to Planes: let it be understood that, in forming such Block, the systematic Equation is always taken along with them.

## IX.

When 2 Planes are said to **intersect in a Line at an infinite distance,** let it be understood that they are parallel.

## X.

When 3 Planes are said to **intersect in a Point at an infinite distance**, let it be understood that either every 2 of them are parallel, or, if any 2 of them intersect, any Line, in which they intersect, is parallel to the 3rd Plane.

---

## Proposition VIII. Th.

If the Equation to a Line, passing through the origin or a vertex of reference, be given in the form

$$\frac{x}{l} = \frac{y}{m} = \frac{z}{n}, \qquad \text{or} \qquad \frac{\alpha}{l} = \frac{\beta}{m} = \frac{\gamma}{n};$$

this may be written in the form

$$\left\| \begin{matrix} x, & y, & z \\ l, & m, & n \end{matrix} \right\| = 0, \qquad \text{or} \qquad \left\| \begin{matrix} \alpha, & \beta, & \gamma \\ l, & m, & n \end{matrix} \right\| = 0.$$ (Ch. V. Prop. IV.

Again, if the Equation to a Line, passing through a given Point, be given in the form

$$\frac{x-x'}{l} = \frac{y-y'}{m} = \frac{z-z'}{n}, \quad \text{or} \quad \frac{\alpha-\alpha'}{l} = \frac{\beta-\beta'}{m} = \frac{\gamma-\gamma'}{n}:$$

this may be written in the form

$$\begin{Vmatrix} x, & y, & z, & 1 \\ x', & y', & z', & 1 \\ l, & m, & n, & 0 \end{Vmatrix} = 0, \quad \text{or} \quad \begin{Vmatrix} \alpha, & \beta, & \gamma, & 1 \\ \alpha', & \beta', & \gamma, & 1 \\ l, & m, & n, & 0 \end{Vmatrix} = 0. \qquad \text{(CH. V. PROP. V.}$$

Again, if the Equation to a Line, passing through 2 given Points, be given in the form

$$\frac{x-x_1}{x_1-x_2} = \frac{y-y_1}{y_1-y_2} = \frac{z-z_1}{z_1-z_2}, \quad \text{or} \quad \frac{\alpha-\alpha_1}{\alpha_1-\alpha_2} = \frac{\beta-\beta_1}{\beta_1-\beta_2} = \frac{\gamma-\gamma_1}{\gamma_1-\gamma_2}:$$

this may be written in the form

$$\begin{Vmatrix} x, & y, & z, & 1 \\ x_1, & y_1, & z_1, & 1 \\ x_2, & y_2, & z_2, & 1 \end{Vmatrix} = 0, \quad \text{or} \quad \begin{Vmatrix} \alpha, & \beta, & \gamma, & 1 \\ \alpha_1, & \beta_1, & \gamma_1, & 1 \\ \alpha_2, & \beta_2, & \gamma_2, & 1 \end{Vmatrix} = 0. \qquad \text{(CH. V. PROP. VI.}$$

## PROPOSITION IX. TH.

If there be given 2 Planes, represented, in the Cartesian System, by

$$A_1x + B_1y + C_1z + D_1 = 0,$$
$$A_2x + B_2y + C_2z + D_2 = 0;$$

or, in the Quadriplanar System, by

$$A_1\alpha + B_1\beta + C_1\gamma + D_1\delta + E_1 = 0,$$
$$A_2\alpha + B_2\beta + C_2\gamma + D_2\delta + E_2 = 0,$$

the systematic Equation being $a\alpha + b\beta + c\gamma + d\delta - 3M = 0$;

and

(1) If $|V| \neq 0$;

then the Equations are consistent, and there is one Variable to which an arbitrary value may be given; (CH. III. PROP. II.

∴ the 2 Planes intersect in more than one Point;

i. e. they intersect in a Line.

also, for each such arbitrary value, there is only one value for each of the other Variables; (CH. III. PROP. II.

∴ they do not coincide;

∴ the 2 Planes intersect in one Line, and in one only.

(2) If $\|V\|=0$, but $|B| \neq 0$;

then the Equations are inconsistent; (CH. III. PROP. V.

∴ the 2 Planes are parallel *;

(3) If $\|B\|=0$; (whence also $\|V\|=0$);

then, in the Cartesian System, the Equations are identical; (CH. III. PROP. VI.

also, in the Quadriplanar System, since either of the Equations to the Planes, taken with the systematic Equation, are such that their $V$-Block is not evanescent; (CONV. VII.

∴ the 3 Equations are consistent, and either of the Equations to the Planes is dependent on the other Equations; (CH. III. PROP. IX. COR.

i. e. whatsoever Point lies on one of the 2 Planes, lies also on the other; (CH. III. DEF. VI.

∴ in either System, the 2 Planes coincide.

From (2) and (3) may be deduced

(4) If $\|V\|=0$;

then the 2 Planes have the same direction.

---

* *Prop. IX.* (2.) Thus, in the Quadriplanar System, if the systematic Equation be

$$9\alpha+9\beta+12\gamma+20\delta-20 = 0;$$

and if there be 2 Planes

$$5\alpha+4\beta+2\gamma+9\delta+5 = 0,$$
$$\alpha-\beta-8\gamma-2\delta+3 = 0:$$

since $\|V\|=0$, (as may be proved by taking the 2 principal Minors of it, which contain the non-evanescent secondary Minor $\left\{\begin{matrix}9, & 9\\ 5, & 4\end{matrix}\right\}$), but $|B| \neq 0$, the 2 Planes are parallel.

N.B. The above systematic Equation was obtained by taking, as the base of the Tetrahedron of reference, a triangle whose sides are 3, 3, 4, and erecting, at the centre of the inscribed, circle a perpendicular whose length $=1$.

## PROPOSITION X. TH.

If there be given 3 Planes, represented, in the Cartesian System, by

$$A_1x+B_1y+C_1z+D_1 = 0,$$
$$A_2x+B_2y+C_2z+D_2 = 0,$$
$$A_3x+B_3y+C_3z+D_3 = 0;$$

or, in the Quadriplanar System, by

$$A_1\alpha+B_1\beta+C_1\gamma+D_1\delta+E_1 = 0,$$
$$A_2\alpha+B_2\beta+C_2\gamma+D_2\delta+E_2 = 0,$$
$$A_3\alpha+B_3\beta+C_3\gamma+D_3\delta+E_3 = 0,$$

the systematic Equation being $a\alpha + b\beta + c\gamma + d\delta - 3M = 0$;

and

(1) If $V \neq 0$;

then the Equations are consistent, and there is only one set of values for the Variables; (CH. III. PROP. I.

∴ the 3 Planes intersect in one Point, and in one only.

(2) If $V=0$, but $|B| \neq 0$;

then the Equations are inconsistent; (CH. III. PROP. IV.

∴ the 3 Planes do not intersect in one Point.

i. e. either every 2 of them are parallel, or, if any 2 of them intersect, any Line, in which they intersect, is parallel to the 3rd Plane;

i. e. they intersect in a Point at an infinite distance.

(3) If $\|B\|=0$; (whence also $V=0$); and if there be, among the Equations to the 3 Planes, 2 which have their $V$-Block not evanescent;

then there are 2 of the Planes which intersect in one Line, and in one only; (PROP. IX. (1)

and the Equation to the 3rd Plane is dependent on the other 3 Equations; (CH. III. PROP. VIII.

∴ the 3rd Plane passes through the line of intersection;

∴ the 3 Planes intersect in one Line, and in one only *.

(4) If $\|B\| = 0$; (whence also $V = 0$); and if every 2 of the Equations to the 3 Planes have their $V$-Block evanescent; and if there be among them 2 which have their $B$-Block not evanescent;

then 2 of the Planes are parallel; (Prop. IX. (2)

and the 3 have the same direction. (Prop. IX. (4)

(5) If every 2 of the Equations to the 3 Planes have their $B$-Block evanescent; (whence also $\|B\| = 0$, and $V = 0$);

then the 3 Planes coincide. (Prop. IX. (3)

From (2), (3), (4), and (5) may be deduced

(6) If $V = 0$;

then the Lines of intersection, if any, have the same direction.

From (3), (4), and (5) may be deduced

(7) If $\|B\| = 0$;

then the 3 Planes intersect in one Line at a finite or infinite distance.

From (3) and (5) may be deduced

(8) If $\|B\| = 0$; and if either there be, among the Equations to the 3 Planes, 2 which have their $V$-Block not evanescent, or if every 2 have their $B$-Block evanescent;

then the 3 Planes intersect in one Line.

---

* *Prop. X.* (3.) Thus the 3 Planes

$$2x + y - z = 0,$$
$$x + 2y + 3z = 0,$$
$$x - 4y - 11z = 0,$$

intersect in one Line, and in one only.

PROPOSITION XI. TH.

If there be given 4 Planes, represented, in the Cartesian System, by

$$A_1x + B_1y + C_1z + D_1 = 0,$$
&c.
$$A_4x + B_4y + C_4z + D_4 = 0;$$

or, in the Quadriplanar System, by

$$A_1\alpha + B_1\beta + C_1\gamma + D_1\delta + E_1 = 0,$$
&c.
$$A_4\alpha + B_4\beta + C_4\gamma + D_4\delta + E_4 = 0;$$

the systematic Equation being $a\alpha + b\beta + c\gamma + d\delta - 3M = 0$;

and

(1) If $B \neq 0$; (whence also $|V| \neq 0$, and every 3 of the Equations to the 4 Planes have their $B$-Block not evanescent);

then the Equations are inconsistent; (CH. III. PROP. III.

∴ the 4 Planes do not intersect in one Point, but there are 3 of them which intersect in one Point, and in one only.

(2) If $B = 0$, and if there be, among the Equations to the 4 Planes, 3 which have their $V$-Block not evanescent;

then the Equations are consistent, and there is only one set of values for the Variables; (CH. III. PROP. VIII.

∴ the 4 Planes intersect in one Point, and in one only *.

---

* *Prop. XI.* (2.) Thus, in the Quadriplanar System, if the systematic Equation be

$$9\alpha + 9\beta + 12\gamma + 20\delta - 20 = 0,$$

the 4 planes

$$\alpha + \beta - \gamma + 5\delta - 3 = 0,$$
$$5\alpha + 3\beta + 8\gamma + 12\delta - 11 = 0,$$
$$3\alpha + 5\beta + 5\gamma + \delta - 4 = 0,$$
$$\alpha + \beta + 3\gamma + 3\delta - 1 = 0,$$

intersect in one Point, and in one only.

(3) If $B = 0$; and if every 3 of the Equations to the 4 Planes have their $V$-Block evanescent; and if there be among them 3 which have their $B$-Block not evanescent;

then 3 of the Planes intersect in a Point at an infinite distance; (Prop. X. (2)

and all the Lines of intersection, if any, have the same direction. (Prop. X. (6)

(4) If every 3 of the Equations to the 4 Planes have their $B$-Block evanescent; (whence also $B = 0$); and if there be among them 2 which have their $V$-Block not evanescent;

then there are 3 of the Planes which intersect in one line, and one only; (Prop. X. (3)

and the Equation of the 4th Plane is dependent on the other Equations; (Ch. III. Prop. IX.

∴ the 4th Plane passes through the Line of intersection;

∴ the 4 Planes intersect in one Line, and one only.

(5) If every 3 of the Equations to the 4 Planes have their $B$-Block evanescent; (whence also $B = 0$); and if every 2 have their $V$-Block evanescent; and if there be among them 2 which have their $B$-Block not evanescent;

then there are 2 of the Planes which are parallel; (Prop. IX. (2)

and the 4 have the same direction. (Prop. IX. (4)

(6) If every 2 of the Equations to the 4 Planes have their $B$-Block evanescent; (whence also every 3 have the same, and $B = 0$);

then the 4 Planes coincide. (Prop. IX. (3)

From (2), (3), (4), (5), and (6) may be deduced

(7) If $B = 0$;

then the 4 Planes intersect in one Point at a finite or infinite distance.

From (2), (4), and (6) may be deduced

(8) If $B = 0$; and if either there be, among the Equations to the 4 Planes, 3 which have their $V$-Block not evanescent, or if every 3 have their $B$-Block evanescent and there be 2 which have their $V$-Block not evanescent, or if every 2 have their $B$-Block evanescent;

then the 4 Planes intersect in one Point.

### PROPOSITION XII. TH.

If there be given 2 Points in Space, represented, in the Quadriplanar System, by $(\alpha_1, \beta_1, \gamma_1, \delta_1)$, $(\alpha_2, \beta_2, \gamma_2, \delta_2)$; and if

$$\begin{Vmatrix} \alpha_1, & \beta_1, & \gamma_1, & \delta_1 \\ \alpha_2, & \beta_2, & \gamma_2, & \delta_2 \end{Vmatrix} = 0;$$

then $$\begin{Vmatrix} \alpha_1, & \beta_1, & \gamma_1, & \delta_1, & 1 \\ \alpha_2, & \beta_2, & \gamma_2, & \delta_2, & 1 \end{Vmatrix} = 0.$$

For $$a\alpha_1 + b\beta_1 + c\gamma_1 + d\delta_1 - 3M = 0,$$
$$a\alpha_2 + b\beta_2 + c\gamma_2 + d\delta_2 - 3M = 0;$$

in these Equations, let $a, b, c, d, -3M$, be considered as Variables;

then there are 2 homogeneous Equations containing 5 Variables; and their $V$-Block is $\left\{ \begin{matrix} \alpha_1, & \beta_1, & \gamma_1, & \delta_1, & 1 \\ \alpha_2, & \beta_2, & \gamma_2, & \delta_2, & 1 \end{matrix} \right\}$; and there is a set of values, for the Variables, of which the last is actual; and, if the last column of their $V$-Block be omitted, the remaining Block is evanescent;

∴ the whole Block is evanescent. (CH. III. PROP. XV. COR.

Therefore, if there be given, &c. Q. E. D.

### COROLLARIES TO PROP. XII.

1.

Hence the 2 Points are coincident.

For, since $\begin{Vmatrix} \alpha_1, & 1 \\ \alpha_2, & 1 \end{Vmatrix} = 0$, $\alpha_1 = \alpha_2$; and so of the others.

2.

By a similar process it may be proved that, if there be 3 Points in Space, thus represented, and if

$$\left\| \begin{matrix} \alpha_1, & \beta_1, & \gamma_1, & \delta_1 \\ \alpha_2, & \beta_2, & \gamma_2, & \delta_2 \\ \alpha_3, & \beta_3, & \gamma_3, & \delta_3 \end{matrix} \right\| = 0 :$$

then
$$\left\| \begin{matrix} \alpha_1, & \beta_1, & \gamma_1, & \delta_1, & 1 \\ \alpha_2, & \beta_2, & \gamma_2, & \delta_2, & 1 \\ \alpha_3, & \beta_3, & \gamma_3, & \delta_3, & 1 \end{matrix} \right\| = 0 :$$

3.

And similarly, (using Ch. III. Prop. XIV. Cor.), that if there be 4 Points in Space, thus represented, and if

$$\left\| \begin{matrix} \alpha_1, & \beta_1, & \gamma_1, & \delta_1 \\ \vdots & \vdots & \vdots & \vdots \\ \alpha_4, & \beta_4, & \gamma_4, & \delta_4 \end{matrix} \right\| = 0 :$$

then
$$\left\| \begin{matrix} \alpha_1, & \beta_1, & \gamma_1, & \delta_1, & 1 \\ \vdots & \vdots & \vdots & \vdots & \vdots \\ \alpha_4, & \beta_4, & \gamma_4, & \delta_4, & 1 \end{matrix} \right\| = 0.$$

## Proposition XIII. Th.

If there be given 3 Points in Space, represented, in the Cartesian System, by $(x_1, y_1, z_1)$, &c.; or, in the Quadriplanar System, by $(\alpha_1, \beta_1, \gamma_1, \delta_1)$, &c.; and if, in each System, the general Equation to a Plane be taken, involving 4 undetermined quantities $A, B, C, D$; and if the coordinates of the given Points be successively substituted in it; and if $A$, $B$, $C$, and $D$ be considered as Variables in the Equations so formed, viz.—

in the Cartesian System,

$$\left. \begin{aligned} Ax_1 + By_1 + Cz_1 + D &= 0, \\ Ax_2 + By_2 + Cz_2 + D &= 0, \\ Ax_3 + By_3 + Cz_3 + D &= 0, \end{aligned} \right\}, \text{ whose } V\text{-Block is } \left\{ \begin{matrix} x_1, & y_1, & z_1, & 1 \\ x_2, & y_2, & z_2, & 1 \\ x_3, & y_3, & z_3, & 1 \end{matrix} \right\};$$

in the Quadriplanar System,

$$\left.\begin{array}{l} A\alpha_1+B\beta_1+C\gamma_1+D\delta_1 = 0, \\ A\alpha_2+B\beta_2+C\gamma_2+D\delta_2 = 0, \\ A\alpha_3+B\beta_3+C\gamma_3+D\delta_3 = 0, \end{array}\right\}, \text{ whose } V\text{-Block is } \left\{\begin{array}{l} \alpha_1, \beta_1, \gamma_1, \delta_1 \\ \alpha_2, \beta_2, \gamma_2, \delta_2 \\ \alpha_3, \beta_3, \gamma_3, \delta_3 \end{array}\right\};$$

then, in the first place, it is evident that there is a real Plane on which the 3 Points lie ;

also further

(1) If $|V| \neq 0$ ;

then the values for the Variables bear to each other one and the same set of ratios ; (CH. III. PROP. II. COR.

∴ there is only one such Plane ;

∴ the 3 Points lie on one Plane, and on one only.

(2) If $\|V\| = 0$ ; and if there be, among the 3 Equations, 2 which have their $V$-Block not evanescent ;

then the Equations are consistent, and there are 2 Variables to which arbitrary values may be given, and for each such set of arbitrary values, there is only one value for each of the other Variables ;

(CH. III. PROP. IX. COR.

∴ for each such set, there is only one Plane ;

hence, by giving to these 2 Variables certain arbitrary values, and again certain others not equimultiples of these, 2 Planes may be found on each of which the 3 Points lie ;

and these Planes do not coincide ; (PROP. IX. (1), (2)

∴ the 3 Points lie on one Line, and on one only.

(3) If every 2 of the 3 Equations have their $V$-Block evanescent ; (whence also $\|V\| = 0$) ;

then the 3 Points coincide. (PROP. XII. COR. 1.

From (2) and (3) may be deduced

(4) If $\|V\| = 0$ ; the 3 Points lie on one Line *.

---

* *Prop. XIII.* (4.) Thus, in the Quadriplanar System, the 3 Points, whose coordinates are

$$\begin{array}{l} (3, \ 1, \ 2, -1), \\ (2, -1, \ 5, \ 4), \\ (1, -3, \ 8, \ 9), \end{array}$$

lie on one Line.

## Proposition XIV. Th.

If there be given 4 Points in Space, represented, in the Cartesian System, by $(x_1, y_1, z_1)$, &c.; or, in the Quadriplanar System, by $(\alpha_1, \beta_1, \gamma_1, \delta_1)$, &c.; and if, in each System, the general Equation to a Plane be taken, involving 4 undetermined quantities $A, B, C, D$; and if the coordinates of the given Points be successively substituted in it; and if the quantities $A, B, C$, and $D$ be considered as Variables in the Equations so formed; viz.—

in the Cartesian System,

$$\left.\begin{array}{c} Ax_1 + By_1 + Cz_1 + D_1 = 0, \\ \text{\&c.} \\ Ax_4 + By_4 + Cz_4 + D_4 = 0, \end{array}\right\}, \text{ whose } V\text{-Block is } \left\{\begin{array}{cccc} x_1, & y_1, & z_1, & 1 \\ \vdots & \vdots & \vdots & \vdots \\ x_4, & y_4, & z_4, & 1 \end{array}\right\};$$

in the Quadriplanar System,

$$\left.\begin{array}{c} A\alpha_1 + B\beta_1 + C\gamma_1 + D\delta_1 = 0, \\ \text{\&c.} \\ A\alpha_4 + B\beta_4 + C\gamma_4 + D\delta_4 = 0, \end{array}\right\}, \text{ whose } V\text{-Block is } \left\{\begin{array}{cccc} \alpha_1, & \beta_1, & \gamma_1, & \delta_1 \\ \vdots & \vdots & \vdots & \vdots \\ \alpha_4, & \beta_4, & \gamma_4, & \delta_4 \end{array}\right\};$$

and

(1) If $V \neq 0$;

then the only values for $A, B, C, D$ are zero; (Ch. III. Prop. I. Cor.

∴ the 4 Points do not lie on one Plane.

(2) If $V = 0$; and if there be, among the 4 Equations, 3 which have their $V$-Block not evanescent;

then there are 3 of the Points which lie on one Plane, and on one only; (Prop. XIII. (1)

also the 4th Equation is dependent on the others; (Ch. III. Prop. IX.

that is, these values of $A, B, C$, and $D$, which belong to a plane passing through these 3 Points, the same belong also to a Plane passing through the $4^{th}$ Point;

∴ the 4 Points lie on one Plane, and on one only *.

(3) If every 3 of the 4 Equations have their $V$-Block evanescent; (whence also $V=0$); and if there be among them 2 which have their $V$-Block not evanescent;

then there are 2 of the Points which do not coincide, that is, which lie on one Line, and on one only;

and each of the other Points lies on the same Line; (PROP. XIII. (2)

∴ the 4 Points lie on one Line, and on one only.

(4) If every 2 of the 4 Equations have their $V$-Block evanescent; (whence also every 3 have their $V$-Block evanescent; and whence also $V=0$);

then the 4 Points coincide. (PROP. XII. COR. 1.

From (2), (3), and (4) may be deduced

(5) If $V=0$; the 4 Points lie on one Plane.

From (3) and (4) may be deduced

(6) If every 3 of the 4 Equations have their $V$-Block evanescent; the 4 Points lie on one Line.

---

* *Prop. XIV.* (2.) Thus, in the Cartesian System, the 4 points, whose coordinates are

$$(2,\ 1,\ -1),$$
$$(3,\ 2,\ 1),$$
$$(1,\ -2,\ -3),$$
$$(4,\ 5,\ 3),$$

lie on one Plane, and on one only.

# CHAPTER VIII.

## *GEOMETRICAL TESTS.*

### SECTION I.

*Plane Geometry.*

#### PROPOSITION I. TH.

*Test for 2 Lines having the same direction.*

If there be given 2 Lines represented, in the Cartesian System, by

$$A_1x + B_1y + C_1 = 0,$$
$$A_2x + B_2y + C_2 = 0;$$

or, in the Trilinear System, by

$$A_1\alpha + B_1\beta + C_1\gamma + D_1 = 0,$$
$$A_2\alpha + B_2\beta + C_2\gamma + D_2 = 0;$$

the systematic Equation being $a\alpha + b\beta + c\gamma - 2M = 0$:

a test for their having the same direction is that $V = 0$.

Let the test be fulfilled;

then the 2 Lines have the same direction; (CH. VII. PROP. III. (4)

∴ the test is *sufficient.*

Next, let it be not fulfilled;
then the 2 Lines intersect in one Point, and in one only;
(CH. VII. PROP. III. (1)
that is, they have not the same direction;
∴ the test is *necessary.*

Therefore, if there be, &c. Q. E. D.

## PROPOSITION II. TH.

*Test for 3 Lines intersecting,* (1) *in a Point at a finite or infinite distance,*
(2) *in one Point.*

If there be given 3 Lines, represented, in the Cartesian System, by

$$A_1x+B_1y+C_1 = 0,$$
$$A_2x+B_2y+C_2 = 0,$$
$$A_3x+B_3y+C_3 = 0;$$

or, in the Trilinear System,

$$A_1\alpha+B_1\beta+C_1\gamma+D_1 = 0,$$
$$A_2\alpha+B_2\beta+C_2\gamma+D_2 = 0,$$
$$A_3\alpha+B_3\beta+C_3\gamma+D_3 = 0;$$

the systematic Equation being $a\alpha + b\beta + c\gamma - 2M = 2$:

then, firstly,

a test for their intersecting in one Point, at a finite or infinite distance, is that $B = 0$.

Let the test be fulfilled;
then the 3 Lines do so intersect; (CH. VII. PROP. IV. (5)
∴ the test is *sufficient.*

Next, let it be not fulfilled;
then the 3 Lines do not intersect in one Point at a finite distance;
(CH. VII. PROP. IV. (1)
also there must be, among their Equations, 2 which have their $V$-Block not evanescent; for otherwise $B$ would $= 0$;

∴ there are, among the 3 Lines, 2 which intersect in one Point and in one only; (CH. VII. PROP. III. (1)

∴ the 3 Lines do not intersect in a Point at an infinite distance;
∴ the test is *necessary*.

Therefore, if there be, &c. Q. E. D.

Secondly,

a test for their intersecting in one Point is that $B = 0$, and that either there are, among the Equations to the 3 Lines, 2 which have their $V$-Block not evanescent, or else every 2 of them have their $B$-Block evanescent.

Let the test be fulfilled;
then the 3 Lines intersect in one Point; (Ch. VII. Prop. IV. (6)
∴ the test is *sufficient*.

Next, let it be not all fulfilled;
then either $|B| \neq 0$, or else every 2 of the Equations to the 3 Lines have their $V$-Block evanescent, and 2 of them have their $B$-Block not evanescent;
in the first case, the 3 Lines do not intersect in one point; (Ch. VII. Prop. IV. (1)
in the second, 2 of them are parallel; (Ch. VII. Prop. III. (2)
∴ in either case, they do not intersect in one Point;
∴ the test is *necessary*.

Therefore, if there be, &c. Q. E. D.

## Proposition III. Th.

*Test for 3 Points lying on one Line.*

If there be given, 3 Points, represented, in the Cartesian System, by $(x_1, y_1)$, &c.; or, in the Trilinear System, by $(\alpha_1, \beta_1, \gamma_1)$, &c.; and if, in each System, the given coordinates be formed into a Block, thus:—

$$\begin{Bmatrix} x_1, & y_1, & 1 \\ x_2, & y_2, & 1 \\ x_3, & y_3, & 1 \end{Bmatrix}, \qquad \begin{Bmatrix} \alpha_1, & \beta_1, & \gamma_1 \\ \alpha_2, & \beta_2, & \gamma_2 \\ \alpha_3, & \beta_3, & \gamma_3 \end{Bmatrix}:$$

a test for the 3 Points lying on one Line is that the Block so formed is evanescent.

In each System let the general Equation to a line be taken, involving 3 undetermined quantities $A$, $B$, $C$; and let the coordinates of the given Points be successively substituted in it; and let $A$, $B$, and $C$ be considered as Variables in the Equations so formed, whose V-Blocks are those given above.

Now let the test be fulfilled;
then the 3 Points lie on one Line; (CH. VII. PROP. VI. (4)
∴ the test is *sufficient*.

Next, let it be not fulfilled;
then they do not lie on one Line; (CH. VII. PROP. VI. (1)
∴ the test is *necessary*.

Therefore, if there be, &c. Q. E. D.

## COROLLARIES TO PROP. III.

1.

If there be given 2 Points, represented, in the Cartesian System, by $(x_1, y_1)$, $(x_2, y_2)$; or, in the Trilinear System, by $(\alpha_1, \beta_1, \gamma_1)$, $(\alpha_2, \beta_2, \gamma_2)$: the Equation to the Line through them is

$$\begin{vmatrix} x, & y, & 1 \\ x_1, & y_1, & 1 \\ x_2, & y_2, & 1 \end{vmatrix} = 0, \quad \text{or} \quad \begin{vmatrix} \alpha, & \beta, & \gamma \\ \alpha_1, & \beta_1, & \gamma_1 \\ \alpha_2, & \beta_2, & \gamma_2 \end{vmatrix} = 0.$$

2.

The Equation in the Cartesian System may also be written

$$\frac{x-x_1}{x_1-x_2} = \frac{y-y_1}{y_1-y_2}.$$ (CH. V. PROP. IX.

3.

The equation in the Trilinear System may also be written

$$\alpha.\begin{vmatrix} \beta_1, & \gamma_1 \\ \beta_2, & \gamma_2 \end{vmatrix} - \beta.\begin{vmatrix} \alpha_1, & \gamma_1 \\ \alpha_2, & \gamma_2 \end{vmatrix} + \gamma.\begin{vmatrix} \alpha_1, & \beta_1 \\ \alpha_2, & \beta_2 \end{vmatrix} = 0.$$

Also, since it is equivalent to

$$\begin{Vmatrix} \alpha, & \beta, & \gamma, & 1 \\ \alpha_1, & \beta_1, & \gamma_1, & 1 \\ \alpha_2, & \beta_2, & \gamma_2, & 1 \end{Vmatrix},$$ (Ch. VII. Prop. V. Cor. 2.

it may be written

$$\frac{\alpha-\alpha_1}{\alpha_1-\alpha_2} = \frac{\beta-\beta_1}{\beta_1-\beta_2} = \frac{\gamma-\gamma_1}{\gamma_1-\gamma_2}.$$ (Ch. V. Prop. IX.

4.

If there be 2 Points, not coincident, represented, in the Trilinear System, by $(\alpha_1, \beta_1, \gamma_1)$, $(\alpha_2, \beta_2, \gamma_2)$: the 3 ratios

$$\begin{vmatrix} \beta_1, & \gamma_1 \\ \beta_2, & \gamma_2 \end{vmatrix} : a, \qquad -\begin{vmatrix} \alpha_1, & \gamma_1 \\ \alpha_2, & \gamma_2 \end{vmatrix} : b, \qquad \begin{vmatrix} \alpha_1, & \beta_1 \\ \alpha_2, & \beta_2 \end{vmatrix} : c,$$

cannot be all equal.

For, if they were, the Equation to the Line through them might be written

$$a\alpha + b\beta + c\gamma = 0;$$

but this does not represent a real Line. (Ch. VII. Prop. I. (2)

# CHAPTER VIII. (*Continued.*)

## SECTION II.

### *Solid Geometry.*

#### PROPOSITION IV. TH.

*Test for 2 Planes having the same direction.*

If there be given 2 Planes, represented, in the Cartesian System, by

$$A_1x + B_1y + C_1z + D_1 = 0,$$
$$A_2x + B_2y + C_2z + D_2 = 0;$$

or, in the Quadriplanar System, by

$$A_1\alpha + B_1\beta + C_1\gamma + D_1\delta + E_1 = 0,$$
$$A_2\alpha + B_2\beta + C_2\gamma + D_2\delta + E_2 = 0;$$

the systematic Equation being $a\alpha + b\beta + c\gamma + d\delta + 3M = 0$;

a test for their having the same direction is that $\| V \| = 0$.

Let the test be fulfilled;
then the 2 Planes have the same direction; (CH. VII. PROP. IX. (4)
∴ it is *sufficient*.

Next, let it be not fulfilled;
then they intersect in one Line and one only; (CH. VII. PROP. IX. (1)

that is, they have not the same direction ;

∴ it is *necessary*.

Therefore, if there be, &c. Q. E. D.

## PROPOSITION V. TH.

*Test for 3 Planes intersecting,* (1) *in one Line at a finite or infinite distance,* (2) *in one Line.*

If there be given 3 Planes, represented, in the Cartesian System, by

$$A_1x + B_1y + C_1z + D_1 = 0,$$
$$A_2x + B_2y + C_2z + D_2 = 0,$$
$$A_3x + B_3y + C_3z + D_3 = 0;$$

or, in the Quadriplanar System, by

$$A_1\alpha + B_1\beta + C_1\gamma + D_1\delta + E_1 = 0,$$
$$A_2\alpha + B_2\beta + C_2\gamma + D_2\delta + E_2 = 0,$$
$$A_3\alpha + B_3\beta + C_3\gamma + D_3\delta + E_3 = 0;$$

the systematic Equation being $a\alpha + b\beta + c\gamma + d\delta - 3M = 0$:

then, firstly,

a test for their intersecting in one Line, at a finite or infinite distance, is that $\|B\| = 0$.

Let the test be fulfilled ;

then the 3 Planes do so intersect ; (CH. VII. PROP. X. (7)

∴ the test is *sufficient*.

Next, let it be not fulfilled ;

then the 3 Planes do not intersect in one Line at a finite distance ; (CH. VII. PROP. X. (1), (2)

also there must be, among their Equations, 2 which have their *V*-Block not evanescent ; for otherwise, every secondary Minor of the *V*-Block would be evanescent, that is, $B$ would $= 0$ ;

∴ there are, among the 3 Planes, 2 which intersect in one Line, and in one only ; (CH. VII. PROP. IX. (1)

∴ the 3 Planes do not intersect in a Line at an infinite distance ;

∴ the test is *necessary*.

Therefore a test, &c. Q. E. D.

Secondly,

a test for their intersecting in one Line is that $\|B\|=0$, and either there are, among the Equations to the 3 Planes, 2 which have their $V$-Block not evanescent, or else every 2 of them have their $B$-Block evanescent.

Let the test be fulfilled;

then the 3 Planes intersect in one Line; (CH. VII. PROP. X. (8)

∴ it is *sufficient*.

Next, let it be not all fulfilled;

then either $|B| \neq 0$; or else every 2 of the Equations to the 3 Planes have their $V$-Block evanescent, and 2 of them have their $B$-Block not evanescent;

in the case where $|B| \neq 0$, the 3 Planes do not intersect in one Line; (CH. VII. PROP. X. (1)

in the second case, 2 of them are parallel; (CH. VII. PROP. IX. (2)

∴ in either case they do not intersect in one Line;

∴ the test is *necessary*.

Therefore, if there be, &c. Q. E. D.

## PROPOSITION VI. TH.

*Test for 4 Planes intersecting,* (1) *in one Point at a finite or infinite distance,* (2) *in one Point.*

If there be given 4 Planes, represented, in the Cartesian System, by

$$A_1x+B_1y+C_1z+D_1 \quad = 0,$$
&c.;

or, in the Quadriplanar System, by

$$A_1\alpha+B_1\beta+C_1\gamma+D_1\delta+E_1 = 0,$$
&c.;

the systematic Equation being $\quad a\alpha+b\beta+c\gamma+d\delta-3M=0$:

then, firstly,

a test for their intersecting in one Point, at a finite or infinite distance, is that $B=0$.

Let the test be fulfilled;
then the 4 Planes do so intersect; (CH. VII. PROP. XI. (7)
∴ the test is *sufficient.*

Next, let it be not fulfilled;
then the 4 Planes do not intersect in one Point at a finite distance; (CH. VII. PROP. XI. (1)
also there must be, among their Equations, 3 which have their *V*-Block not evanescent; for otherwise *B* would $= 0$;
∴ there are, among the 4 Planes, 3 which intersect in one Point and in one only; (CH. VII. PROP. X. (1)
∴ the 4 Planes do not intersect in a Point at an infinite distance;
∴ the test is *necessary.*

Therefore a test, &c. Q. E. D.

Secondly,

a test for their intersecting in one Point is that $B = 0$, and either there are, among the Equations to the 4 Planes, 3 which have their *V*-Block not evanescent, or else every 3 have their *B*-Block evanescent and there are 2 which have their *V*-Block not evanescent, or else every 2 have their *B*-Block evanescent.

Let the test be fulfilled;
then the 4 Planes intersect in one Point. (CH. VII. PROP. XI. (8)
∴ it is *sufficient.*

Next, let it be not all fulfilled;
then either $B \neq 0$; or every 3 of the Equations to the 4 Planes have their *V*-Block evanescent, and there are 3 among them which have their *B*-Block not evanescent; or every 2 have their *V*-Block evanescent, and there are 2 which have their *B*-Block not evanescent;
in the first case, the 4 Planes do not intersect in one Point; (CH. VII. PROP. XI. (1)
in the second, 3 of them do not so intersect; (CH. VII. PROP. X. (2)
in the third, 2 of them are parallel; (CH. VII. PROP. IX. (2)
∴ in any case, they do not intersect in one Point;
∴ the test is *necessary.*

Therefore, if there be, &c. Q. E. D.

## PROPOSITION VII. TH.

*Test for 3 Points in Space lying on one Line.*

If there be given 3 Points in Space, represented, in the Cartesian System, by $(x_1, y_1, z_1)$, &c.; or, in the Quadriplanar System, by $(\alpha_1, \beta_1, \gamma_1, \delta_1)$, &c.; and if, in each System, the given coordinates be formed into a Block, thus :—

$$\left\{\begin{matrix} x_1, & y_1, & z_1, & 1 \\ x_2, & y_2, & z_2, & 1 \\ x_3, & y_3, & z_3, & 1 \end{matrix}\right\}, \qquad \left\{\begin{matrix} \alpha_1, & \beta_1, & \gamma_1, & \delta_1 \\ \alpha_2, & \beta_2, & \gamma_2, & \delta_2 \\ \alpha_3, & \beta_3, & \gamma_3, & \delta_3 \end{matrix}\right\}:$$

a test for the 3 Points lying on one Line is that the Block so formed is evanescent.

In each System let the general equation to a Plane be taken, involving 4 undetermined quantities *A, B, C, D*; and let the coordinates of the given Points be successively substituted in this general Equation; and let *A, B, C,* and *D* be considered as Variables in the Equations so formed, so that their *V*-Blocks are those given above.

Now let the test be fulfilled;
then the 3 Points lie on one Line; (CH. VII. PROP. XIII. (4)
∴ it is *sufficient*.

Next, let it be not fulfilled;
then the 3 Points lie on one Plane, and one only;
(CH. VII. PROP. XIII. (1)
∴ it is *necessary*.

Therefore, if there be, &c. Q. E. D.

### COROLLARIES TO PROP. VII.

1.

If there be given 2 Points in Space, represented, in the Cartesian System, by $(x_1, y_1, z_1)$, &c.; or, in the Quadriplanar System, by $(\alpha_1, \beta_1, \gamma_1, \delta_1)$, &c.: the Equations to the Line through them are given by

$$\left\|\begin{matrix} x, & y, & z, & 1 \\ x_1, & y_1, & z_1, & 1 \\ x_2, & y_2, & z_2, & 1 \end{matrix}\right\| = 0, \qquad \left\|\begin{matrix} \alpha, & \beta, & \gamma, & \delta \\ \alpha_1, & \beta_1, & \gamma_1, & \delta_1 \\ \alpha_2, & \beta_2, & \gamma_2, & \delta_2 \end{matrix}\right\| = 0.$$

2.

The Equation in the Cartesian System may also be written

$$\frac{x-x_1}{x_1-x_2}=\frac{y-y_1}{y_1-y_2}=\frac{z+z_1}{z_1-z_2}. \qquad \text{(Ch. V. Prop. IX.}$$

3.

The Equation in the Quadriplanar System is equivalent to

$$\left\|\begin{matrix}\alpha, & \beta, & \gamma, & \delta, & 1\\ \alpha_1, & \beta_1, & \gamma_1, & \delta_1, & 1\\ \alpha_2, & \beta_2, & \gamma_2, & \delta_2, & 1\end{matrix}\right\|=0, \qquad \text{(Ch. VII. Prop. V. Cor. 2.}$$

and therefore may be written

$$\frac{\alpha-\alpha_1}{\alpha_1-\alpha_2}=\frac{\beta-\beta_1}{\beta_1-\beta_2}=\frac{\gamma-\gamma_1}{\gamma_1-\gamma_2}=\frac{\delta-\delta_1}{\delta_1-\delta_2}. \qquad \text{(Ch. V. Prop. IX.}$$

## Proposition VIII. Th.

*Test for 4 Points in Space lying on one Plane.*

If there be given 4 Points in Space, represented, in the Cartesian System, by $(x_1, y_1, z_1)$, &c.; or, in the Quadriplanar System, by $(\alpha_1, \beta_1, \gamma_1, \delta_1)$, &c.; and if, in each System, the given coordinates be formed into a Block, thus :—

$$\left\{\begin{matrix}x_1, & y_1, & z_1, & 1\\ x_2, & y_2, & z_2, & 1\\ x_3, & y_3, & z_3, & 1\\ x_4, & y_4, & z_4, & 1\end{matrix}\right\}, \qquad \left\{\begin{matrix}\alpha_1, & \beta_1, & \gamma_1, & \delta_1\\ \alpha_2, & \beta_2, & \gamma_2, & \delta_2\\ \alpha_3, & \beta_3, & \gamma_3, & \delta_3\\ \alpha_4, & \beta_4, & \gamma_4, & \delta_4\end{matrix}\right\}:$$

the test for the 4 Points lying on one Plane is that the Block so formed is evanescent.

In each System let the general Equation to a Plane be taken, involving 4 undetermined quantities $A, B, C, D$; and let the coordinates of the given Points be successively substituted in this general Equation; and let $A, B, C$, and $D$ be considered as Variables in the Equations so formed, so that their $V$-Blocks are those given above.

Now let the test be fulfilled;
then the 4 Points lie on one Plane; (CH. VII. PROP. XIV. (5)
$\therefore$ it is *sufficient.*

Next, let it be not fulfilled;
then the 4 Points do not lie on one Plane; (CH. VII. PROP. XIV. (1)
$\therefore$ it is *necessary.*

Therefore, if there be, &c. Q. E. D.

## COROLLARIES TO PROP. VIII.

1.

If there be given 3 Points in space, represented, in the Cartesian System, by $(x_1, y_1, z_1)$, &c.; or, in the Quadriplanar System, by $(\alpha_1, \beta_1, \gamma_1, \delta_1)$, &c.: the Equation to the Plane through them is

$$\begin{vmatrix} x, & y, & z, & 1 \\ x_1, & y_1, & z_1, & 1 \\ x_2, & y_2, & z_2, & 1 \\ x_3, & y_3, & z_3, & 1 \end{vmatrix} = 0, \quad \text{or} \quad \begin{vmatrix} \alpha, & \beta, & \gamma, & \delta \\ \alpha_1, & \beta_1, & \gamma_1, & \delta_1 \\ \alpha_2, & \beta_2, & \gamma_2, & \delta_2 \\ \alpha_3, & \beta_3, & \gamma_3, & \delta_3 \end{vmatrix} = 0.$$

2.

The Equation in the Quadriplanar System may be written

$$\alpha . \begin{vmatrix} \beta_1, & \gamma_1, & \delta_1 \\ \beta_2, & \gamma_2, & \delta_2 \\ \beta_3, & \gamma_3, & \delta_3 \end{vmatrix} - \beta . \begin{vmatrix} \alpha_1, & \gamma_1, & \delta_1 \\ \alpha_2, & \gamma_2, & \delta_2 \\ \alpha_3, & \gamma_3, & \delta_3 \end{vmatrix} + \gamma . \begin{vmatrix} \alpha_1, & \beta_1, & \delta_1 \\ \alpha_2, & \beta_2, & \delta_2 \\ \alpha_3, & \beta_3, & \delta_3 \end{vmatrix} - \delta . \begin{vmatrix} \alpha_1, & \beta_1, & \gamma_1 \\ \alpha_2, & \beta_2, & \gamma_2 \\ \alpha_3, & \beta_3, & \gamma_3 \end{vmatrix} = 0.$$

3.

If there be 3 Points in Space, not lying in one Line, represented in the Quadriplanar System by $(\alpha_1, \beta_1, \gamma_1, \delta_1)$, &c.; then the ratios

$$\begin{vmatrix} \beta_1, & \gamma_1, & \delta_1 \\ \beta_2, & \gamma_2, & \delta_2 \\ \beta_3, & \gamma_3, & \delta_3 \end{vmatrix} : a, \quad - \begin{vmatrix} \alpha_1, & \gamma_1, & \delta_1 \\ \alpha_2, & \gamma_2, & \delta_2 \\ \alpha_3, & \gamma_3, & \delta_3 \end{vmatrix} : b, \quad \begin{vmatrix} \alpha_1, & \beta_1, & \delta_1 \\ \alpha_2, & \beta_2, & \delta_2 \\ \alpha_3, & \beta_3, & \delta_3 \end{vmatrix} : c, \quad - \begin{vmatrix} \alpha_1, & \beta_1, & \gamma_1 \\ \alpha_2, & \beta_2, & \gamma_2 \\ \alpha_3, & \beta_3, & \gamma_3 \end{vmatrix} : d,$$

cannot be all equal.

For, if they were, the Equation to the Plane through these 3 Points might be written

$$a\alpha + b\beta + c\gamma + d\delta = 0;$$

but this does not represent a real Plane. (CH. VII. PROP. VII. (2)

# APPENDIX I.

## METHOD OF ANALYSING A GIVEN SET OF SIMULTANEOUS LINEAR EQUATIONS. (*See Page* 74, *note.*)

### 1. *Equations not all homogeneous.*

The points, on which information is required, concern

(1) The consistency of the Equations.

(2) Their dependence one on another.

(3) The Variables to which arbitrary values may be simultaneously assigned.

We begin by examining the *V*-Block and *B*-Block of the first 2 Equations; then those of the first 3; of the first 4, and so on.

If in the course of this process we find a set of Equations whose *V*-Block is evanescent, but not their *B*-Block, these are inconsistent, and the inquiry comes to an end.

If in its course we find a set whose *B*-Block is evanescent, the last may be set aside as dependent on one or more of the preceding.

This process is continued until the whole set have been thus examined, or until we have found a set, whose number is equal to the number of the Variables, and whose *V*-Block is not evanescent. In the latter case, if there be Equations still remaining, we must take each of them separately along with the set already examined, and examine the *B*-Block of each set so formed. If any such *B*-Block be not evanescent, the Equations are inconsistent; but if every such *B*-Block be evanescent, all these remaining Equations are dependent on the set already examined.

Thus in any case we either prove the inconsistency of the given Equations, or else (setting aside all that are proved to be dependent on others) we obtain a set of *independent* Equations, whose *V*-Block is not evanescent.

Now in this set of independent Equations, the number of Variables is either equal to, or else greater than, the number of Equations. In the former case, there is only one set of values for the Variables; in the latter, the excess gives the number of Variables to which arbitrary values may be simultaneously assigned, and, for every non-evanescent principal Minor of the *V*-Block, there is such a set of Variables, namely those whose coefficients do not enter into that principal Minor. (Hence, in this case, there are always two such sets at least. See CHAP. V. PROP. X.)

Let us take as an example the 4 Equations

$$\begin{aligned} u+\ v-2x+y-\ z-6&=0,\\ 2u+2v-4x-y+\ z-9&=0,\\ u+\ v-2x\qquad\quad -5&=0,\\ u-\ v+\ x+y-2z\quad &=0. \end{aligned}$$

We begin by examining the *V*-Block and *B*-Block of the first 2 Equations; and for this purpose we take the first column along with each of the others successively. Thus we have $\begin{vmatrix}1, & 1\\ 2, & 2\end{vmatrix}=0,\ \begin{vmatrix}1, & -2\\ 2, & -4\end{vmatrix}=0,\ \begin{vmatrix}1, & 1\\ 2, & -1\end{vmatrix}\neq 0.$ This shows that their *V*-Block is not evanescent.

We now take the first 3 Equations, and combine the 2 columns, which contain the non-evanescent Minor so found, with each of the other columns successively. Thus we have

$$\begin{vmatrix}1, & 1, & 1\\ 2, & -1, & 2\\ 1, & 0, & 1\end{vmatrix}=0,\quad \begin{vmatrix}1, & 1, & -2\\ 2, & -1, & -4\\ 1, & 0, & -2\end{vmatrix}=0,$$

$$\begin{vmatrix}1, & 1, & -1\\ 2, & -1, & 1\\ 1, & 0, & 0\end{vmatrix}=0,\quad \begin{vmatrix}1, & 1, & -6\\ 2, & -1, & -9\\ 1, & 0, & -5\end{vmatrix}=0.$$

This shows that the *B*-Block of these 3 Equations is evanescent, so that the 3rd is dependent on one or both of the first 2.

Omitting the 3rd we take the 1st, 2nd, and 4th, and proceed as before.

Thus we have $\begin{vmatrix} 1, & 1, & 1 \\ 2, & -1, & 2 \\ 1, & 1, & -1 \end{vmatrix} \neq 0$. This shows that the $V$-Block of these 3 Equations is not evanescent. Hence these Equations are consistent, and, since they contain 5 Variables, there are 2 Variables to which arbitrary values may be assigned.

To ascertain how many such sets of 2 may be selected from the 5 Variables, it is necessary to compute *all* the principal Minors of the $V$-Block of these 3 Equations. These are $\begin{vmatrix} 1, & 1, & -2 \\ 2, & 2, & -4 \\ 1, & -1, & 1 \end{vmatrix} = 0$, (because the oblong Block, formed of the first 2 rows, is evanescent); next, taking columns (124) $\begin{vmatrix} 1, & 1, & 1 \\ 2, & 2, & -1 \\ 1, & -1, & 1 \end{vmatrix} \neq 0$; for columns (125), $\begin{vmatrix} 1, & 1, & -6 \\ 2, & 2, & -9 \\ 1, & -1, & 0 \end{vmatrix} \neq 0$; for (134), $\begin{vmatrix} 1, & -2, & 1 \\ 2, & -4, & -1 \\ 1, & 1, & 1 \end{vmatrix} \neq 0$; for (135), $\begin{vmatrix} 1, & -2, & -6 \\ 2, & -4, & -9 \\ 1, & 1, & 0 \end{vmatrix} \neq 0$; for (145), $\begin{vmatrix} 1, & -1, & -6 \\ 2, & 1, & -9 \\ 1, & -2, & 0 \end{vmatrix} \neq 0$; for (234), $\begin{vmatrix} 1, & -2, & -1 \\ 2, & -4, & 1 \\ -1, & 1, & -2 \end{vmatrix} \neq 0$; for (235), $\begin{vmatrix} 1, & -2, & -6 \\ 2, & -4, & -9 \\ -1, & 1, & 0 \end{vmatrix} \neq 0$; for (245), $\begin{vmatrix} 1, & 1, & -6 \\ 2, & -1, & -9 \\ -1, & 1, & 0 \end{vmatrix} \neq 0$; for (345), $\begin{vmatrix} 1, & -1, & -6 \\ -1, & 1, & -9 \\ 1, & -2, & 0 \end{vmatrix} \neq 0$.

We have thus ascertained, with regard to these 4 Equations, that

(1) They are consistent.

(2) The 1st, 2nd, and 4th are independent, and the 3rd is dependent on one or both of the first two.

(3) It is possible to assign arbitrary values to 2 of the Variables simultaneously; and for this purpose *any* set of 2, with the exception of $(y, z)$, may be taken. If, for example, we assign to $v$ and $y$ the values '1', '2', we obtain for $u$, $x$, and $z$ the values '2', '$-1$', '1'.

Again, let us take the 5 Equations

$$3x - y + 7 = 0,$$
$$6x - 2y + 14 = 0,$$
$$x + y + 1 = 0,$$
$$x + 5y - 3 = 0,$$
$$5x + y + 9 = 0.$$

We begin by examining the $V$-Block and $B$-Block of the first 2 Equations; and for this purpose we take the first column along with each of the others successively. Thus we have $\begin{vmatrix} 3, & -1 \\ 6, & -2 \end{vmatrix} = 0$, $\begin{vmatrix} 3, & 7 \\ 6, & 14 \end{vmatrix} = 0$. Hence the $B$-Block is evanescent, and either Equation is dependent on the other.

Omitting the 2nd, we take the 1st and 3rd, and examine them in the same manner. Thus we have $\begin{vmatrix} 3, & -1 \\ 1, & 1 \end{vmatrix} \neq 0$. Hence these Equations are consistent, and there is only one set of values for the Variables.

We have now applied the general process as far as it will go, since the number of Equations, last tested, is equal to the number of Variables. All we have now to do, is to take these 2 Equations along with each of the remaining Equations successively, and examine whether the $B$-Block of each set, so formed, is evanescent or not.

Taking them along with the 4th, we have $\begin{vmatrix} 3, & -1, & 7 \\ 1, & 1, & 1 \\ 1, & 5, & -3 \end{vmatrix} = 0$. Hence the 4th Equation is dependent on one or both of the 1st and 3rd.

Then taking them along with the 5th, we have $\begin{vmatrix} 3, & -1, & 7 \\ 1, & 5, & -3 \\ 5, & 1, & 9 \end{vmatrix} = 0$. Hence the 5th Equation is similarly dependent.

We have thus ascertained, with regard to these 5 Equations, that

(1) They are consistent.

(2) The 1st and 3rd are independent, and each of the rest is dependent on one or both of these.

(3) There are no Variables to which arbitrary values can be assigned.

## 2. *Given Equations all homogeneous.*

It must be remembered that homogeneous Equations are *always* consistent, i.e. they may be satisfied by assigning to each Variable the value zero: n some cases, the Variables admit of no other values, in others, they admit of ı set of values of which one at least is (and therefore two at least are) actual. n this latter case, whatever set of values for the Variables satisfy the Equaions, any equimultiples of them will do so also, so that in this case it is ılways possible to assign an arbitrary value to any one of those Variables vhich admit of actual values. Again, there are cases in which it is possible o assign arbitrary values to 2 or more of these Variables simultaneously. These properties will be the subject of our inquiry.

The points, on which information is required, concern

(1) The possibility of assigning to the Variables a set of values which are not all zero. (In which case 2 at least of the Variables admit of actual values, and to either of them an arbitrary value may be assigned.)

(2) The dependence of the Equations one on another.

(3) The Variables to which arbitrary values may be simultaneously assigned.

We begin by examining the *V*-Block of the first 2 Equations; then those of the first 3; of the first 4, and so on.

If in the course of this process we find a set of Equations whose *V*-Block is evanescent, the last may be set aside as dependent on one or more of the preceding.

This process is continued until the whole set have been thus examined, or until we have found a set, whose number is less by unity than the number of the Variables, and whose *V*-Block is not evanescent. In the latter case, if there be Equations still remaining, we must take each of them separately along with the set already examined, and examine the *V*-Block of each set so formed. If any such *V*-Block be not evanescent, the Equations admit of zero values only; but if every such *V*-Block be not evanescent, all these remaining Equations are dependent on the set already examined.

Thus in any case we either prove that the given equations admit of zero values only, or else (setting aside all that are proved to be dependent on others) we obtain a set of *independent* Equations, whose *V*-Block is not evanescent.

Now in this set of independent Equations, the number of Variables exceeds the number of Equations, either by unity, or by some greater number. In the former case, there is only one set of ratios among the Variables, (see CHAP. III. PROP. II. COR.), i. e. it is not possible to assign arbitrary values to 2 of the Variables simultaneously; in the latter case, the excess gives the number of Variables to which arbitrary values may be simultaneously assigned, and, for every non-evanescent principal Minor of the *V*-Block, there is such a set of Variables, namely those whose coefficients do not enter into that principal Minor. (Hence, in this case, there are always 2 such sets at least. See CHAP. V. PROP. X.)

Let us take as an instance the 3 Equations

$$2u+v+2x+y+3z=0,$$
$$5u+3v-4x+3y-6z=0,$$
$$u+v-8x+y-12z=0.$$

Here the *V*-Block of the first 2 Equations is not evanescent.

But the *V*-Block of the whole set is evanescent; hence the 3rd Equation may be omitted as dependent on the other 2.

And, in these 2, the Variables exceed the Equations in number by 3; hence every non-evanescent principal Minor indicates a set of 3 Variables to which arbitrary values may be simultaneously assigned. The non-evanescent Minors are those belonging to $(u, v)$, $(u, x)$, $(u, y)$, $(u, z)$, $(v, x)$, $(v, z)$, $(x, y)$, $(y, z)$. Hence, with the exception of $(u, x, z)$, $(u, v, y)$, *any* 3 of the Variables may have arbitrary values assigned to them. If, for example, we assign to $x, y, z$, the values 1, 2, $-1$, we obtain for $u$ and $v$ the values 5, $-11$.

# APPENDIX II.

## ARITHMETICAL COMPUTATION OF DETERMINANTS.

A *general* method for computing the value of a Determinant has been already given (see CH. II. PROP. I. COR. 1. Note); and, when the Elements are *Algebraical,* this method is perhaps the best we can employ.

But when the Elements are *Arithmetical,* it is often possible so to re-arrange, or otherwise modify, the given Block, as to make the process of computation both easier and more expeditious. Were not this the case, it would seldom be worth while to employ Determinants for any purpose where actual calculation is necessary; for instance, in the solution of a set of 3 or more simultaneous Equations, the old method of elimination would be far preferable.

In this process of simplification, much must be left to the ingenuity of the student, guided by the circumstances of the case. A few general rules are all that the teacher can supply.

1. We have seen (CH. II. AX. II.) that "*if, in a square Block, the Elements of any one row or column be multiplied by* v; *the Determinant of the new Block is equal to that of the first multiplied by* v." Hence conversely, if, in a square Block, the Elements of any one row, or column, contain $v$ as a factor; it may be divided out and placed outside the Determinant. As an instance of the application of this principle, let us take the Block $\begin{Bmatrix} 36, & 15, & -24 \\ 18, & -9, & 3 \\ 36, & 8, & 16 \end{Bmatrix}$. Here we may observe that the 1st row contains '3' as a factor, the 2nd '3' also, and that the 3rd contains '4': hence the Determinant may be reduced to the form $3^2.4.\begin{vmatrix} 12, & 5, & -8 \\ 6, & -3, & 1 \\ 9, & 2, & 4 \end{vmatrix}$: and this again, since the 1st column contains '3' as a factor, may be reduced to the form $3^3.4.\begin{vmatrix} 4, & 5, & -8 \\ 2, & -3, & 1 \\ 3, & 2, & 4 \end{vmatrix}$

2. We have seen (Ch. II. Prop. I. Cor. 2.) that "*if, in a square Block, the Elements in any one row, or column, all vanish but one: the Determinant of the Block is the product produced by multiplying the Determinant of the complemental Minor of that Element by that Element itself, affected with + or −, according as the numerals in its symbol are similar or dissimilar.*" As an instance of the application of this principle, let us take the Block $\left\{\begin{matrix} 3, & 1, & 0, & 2 \\ 2, & -1, & 0, & 1 \\ 1, & 1, & -5, & 2 \\ 4, & -2, & 0, & -1 \end{matrix}\right\}$, where the Elements of the 3rd column all vanish but one, and where the symbol of that Element is $3\backslash 3$, so that its numerals are *similar*. Hence the Determinant of this Block may be at once reduced to the form $-5.\begin{vmatrix} 3, & 1, & 2 \\ 2, & -1, & 1 \\ 4, & -2, & -1 \end{vmatrix}$. When the given Block does not contain any such row or column, it may be made to do so by the application of another principle, which we proceed to consider.

3. We have seen (Ch. II. Prop. III. Cor. 3.) that "*if, in a square Block, there be added to the several Elements of any row, or column, the corresponding Elements of any other row, or column, multiplied by any number: the Determinant of the new Block is the same as that of the first.*" As an instance of the application of this principle, let us take the Block $\left\{\begin{matrix} 7, & 2, & 1, & -3 \\ 2, & -1, & 3, & 6 \\ 1, & 4, & -3, & 2 \\ -6, & 4, & 1, & 5 \end{matrix}\right\}$, and let us select the 3rd column as the one to be reduced to the required form, and its first Element, "1," as the one which is not to vanish. Now to the Elements of the 2nd row add those of the 3rd. To the Elements of the 3rd row add those of the 4th, multiplied by 3. To the Elements of the 4th row add those of the 1st, multiplied by −1. The Determinant of the Block is thus reduced to the form $\begin{vmatrix} 7, & 2, & 1, & -3 \\ 3, & 3, & 0, & 8 \\ -17, & 10, & 0, & 17 \\ -13, & 2, & 0, & 8 \end{vmatrix}$, which again is reduced, by the former rule, to $\begin{vmatrix} 3, & 3, & 8 \\ -17, & 10, & 17 \\ -13, & 2, & 8 \end{vmatrix}$.

In employing this Rule, we must observe that each modification is a *separate* application of the principle, so that if we employ, in any stage of the process, the Elements of a row, or column, which has been already modified, we must employ them as so modified, and not in their original state. Failing to observe this, we might imagine that, in the Block $\left\{\begin{matrix} 3, & 5, & 2 \\ 2, & 1, & 3 \\ 4, & -1, & -2 \end{matrix}\right\}$, it would be legitimate to add to the Elements of the $2^{nd}$ row those of the $3^{rd}$, and at the same time to add to the Elements of the $3^{rd}$ row those of the $2^{nd}$, and thus to reduce the Determinant to the form $\begin{vmatrix} 3, & 5, & 2 \\ 6, & 0, & 1 \\ 6, & 0, & 1 \end{vmatrix}$: whereas the first modification reduces the Determinant to the form $\begin{vmatrix} 3, & 5, & 2 \\ 6, & 0, & 1 \\ 4, & -1, & -2 \end{vmatrix}$, and thus the second proposed modification is impossible. To guard against this error it is always best to employ, in each stage of the process, the Elements of some row, or column, which has not yet been modified.

4. The process of computation, which I now proceed to explain, and for which "Condensation" appears to be an appropriate name, was communicated by me to the Royal Society in the year 1866, and an account of it is to be found in their "Proceedings," No. 84.

In the following remarks I shall use the phrase "interior of a Block" to denote the Block which remains when the first and last rows and columns are erased.

The process of "Condensation" is exhibited in the following rules, in which the given block is supposed to consist of $n$ rows and $n$ columns:—

(1) Arrange the given Block, if necessary, so that no ciphers occur in its interior. This may be done either by transposing rows or columns, or by adding to certain rows the several terms of other rows multiplied by certain multipliers.

(2) Compute the Determinant of every Minor consisting of four adjacent terms. These values will constitute a second Block, consisting of $\overline{n-1}$ rows and $\overline{n-1}$ columns.

(3) Condense this second Block in the same manner, dividing each term, when found, by the corresponding term in the interior of the first Block.

(4) Repeat this process as often as may be necessary (observing that in condensing any Block of the series, the $r^{\text{th}}$ for example, the terms so found must be divided by the corresponding terms in the interior of the $\overline{r-1}^{\text{th}}$ Block), until the Block is condensed to a single term, which will be the required value.

As an instance of the foregoing rules, let us take the Block

$$\begin{vmatrix} -2 & -1 & -1 & -2 \\ -1 & -2 & -1 & -3 \\ -1 & -1 & 2 & 2 \\ 2 & 1 & -3 & -4 \end{vmatrix}.$$

By rule (2) this is condensed into $\begin{vmatrix} 3 & -1 & 1 \\ -1 & -5 & 4 \\ 1 & 1 & -2 \end{vmatrix}$; this, again, by rule (3), is condensed into $\begin{vmatrix} 8 & -1 \\ -4 & 3 \end{vmatrix}$; and this, by rule (4), into $-4$, which is the required value.

The simplest method of working this rule appears to be to arrange the series of Blocks one under another, as here exhibited; it will then be found very easy to pick out the divisors required in rules (3) and (4).

$$\begin{vmatrix} -2 & -1 & -1 & -2 \\ -1 & -2 & -1 & -3 \\ -1 & -1 & 2 & 2 \\ 2 & 1 & -3 & -4 \end{vmatrix}$$

$$\begin{vmatrix} 3 & -1 & 1 \\ -1 & -5 & 4 \\ 1 & 1 & -2 \end{vmatrix}$$

$$\begin{vmatrix} 8 & -1 \\ -4 & 3 \end{vmatrix}$$

$$-4.$$

This process cannot be continued when ciphers occur in the interior of any one of the Blocks, since infinite values would be introduced by employing them as divisors. When they occur in the given Block itself, it may be re-arranged as has been already mentioned; but this cannot be done when they occur in any one of the derived Blocks; in such a case the given Block must be rearranged as circumstances require, and the operation commenced anew.

The best way of doing this is as follows :—

Suppose a cipher to occur in the $h^{th}$ row and $k^{th}$ column of one of the derived Blocks (reckoning both row and column from the *nearest* corner of the Block); find the term in the $h^{th}$ row and $k^{th}$ column of the given Block (reckoning from the corresponding corner), and transpose rows or columns cyclically until it is left in an outside row or column. When the necessary alterations have been made in the derived Blocks, it will be found that the cipher now occurs in an outside row or column, and therefore need no longer be used as a divisor.

The advantage of *cyclical* transposition is, that most of the terms in the new Blocks will have been computed already, and need only be copied; in no case will it be necessary to compute more than *one* new row or column for each Block of the series. We must of course observe, in any such transposition, whether or no the *sign* of the Determinant is changed.

In the following instance it will be seen that in the first series of Blocks a cipher occurs in the interior of the third. We therefore abandon the process at that point and begin again, re-arranging the given Block by transferring the top row to the bottom; and the cipher, when it occurs, is now found in an exterior row. It will be observed that in each Block of the new series, there is only *one* new row to be computed; the other rows are simply copied from the work already done.

$$\begin{vmatrix} 2 & -1 & 2 & 1 & -3 \\ 1 & 2 & 1 & -1 & 2 \\ 1 & -1 & -2 & -1 & -1 \\ 2 & 1 & -1 & -2 & -1 \\ 1 & -2 & -1 & -1 & 2 \end{vmatrix}$$

$$\begin{vmatrix} 5 & -5 & -3 & -1 \\ -3 & -3 & -3 & 3 \\ 3 & 3 & 3 & -1 \\ -5 & -3 & -1 & -5 \end{vmatrix}$$

$$\begin{vmatrix} -30 & 6 & -12 \\ 0 & 0 & 6 \\ 6 & -6 & 8 \end{vmatrix}$$

$$\begin{vmatrix} 1 & 2 & 1 & -1 & 2 \\ 1 & -1 & -2 & -1 & -1 \\ 2 & 1 & -1 & -2 & -1 \\ 1 & -2 & -1 & -1 & 2 \\ 2 & -1 & 2 & 1 & -3 \end{vmatrix}$$

$$\begin{vmatrix} -3 & -3 & -3 & 3 \\ 3 & 3 & 3 & -1 \\ -5 & -3 & -1 & -5 \\ 3 & -5 & 1 & 1 \end{vmatrix}$$

$$\begin{vmatrix} 0 & 0 & 6 \\ 6 & -6 & 8 \\ -17 & 8 & -4 \end{vmatrix}$$

$$\begin{vmatrix} 0 & 12 \\ 18 & 40 \end{vmatrix}$$

36.

The fact that, whenever ciphers occur in the interior of a derived Block, it is necessary to recommence the operation, may be thought a great obstacle to the use of this method; but I believe it will be found in practice that, even though this should occur several times in the course of one operation, the whole amount of labour will still be much less than that involved in the old process of computation.

# APPENDIX III.

## ALGEBRAICAL PROOF OF THE METHOD OF 'CONDENSATION.'

We have seen (CH. II. PROP. VII.) that "*if there be a square Block of the* $n^{th}$ *degree, and if in it any Minor of the* $m^{th}$ *degree be selected: the Determinant of the corresponding Minor in the adjugate Block is equal, in absolute magnitude, to the product of the* $\overline{m-1}^{th}$ *power of the Determinant of the first Block, multiplied by the Determinant of the Minor complemental to the one selected. Also, if the numerals, indicating the selected rows, be represented by* $\alpha$, $\beta$,......, *and those indicating the selected columns by* $\kappa$, $\lambda$,......, *and their respective sums by* $\Sigma(\alpha)$, $\Sigma(\kappa)$: *the relationship of sign between the equal magnitudes will be secured by multiplying either of them by* $(-1)^{m.(\Sigma(\alpha)+\Sigma(\kappa))}$.

Let us first take a Block of 9 terms, and represent it by

$$\left\{\begin{matrix} 1\backslash 1 \ldots\ldots 1\backslash 3 \\ \vdots \qquad\quad \vdots \\ 3\backslash 1 \ldots\ldots 3\backslash 3 \end{matrix}\right\}_a, \text{ and the adjugate Block by } \left\{\begin{matrix} 1\backslash 1 \ldots\ldots 1\backslash 3 \\ \vdots \qquad\quad \vdots \\ 3\backslash 1 \ldots\ldots 3\backslash 3 \end{matrix}\right\}_A.$$

If we 'condense' this, by the method already given, we get the Block $\left\{\begin{matrix} 3\backslash 3_A, -3\backslash 1_A \\ -1\backslash 3_A, \ 1\backslash 1_A \end{matrix}\right\}$, and the Determinant of this will remain unchanged if we transpose the columns, and also the rows, and then multiply the first row and first column by $-1$;

$$\text{hence it} = \left|\begin{matrix} 1\backslash 1_A, 1\backslash 3_A \\ 3\backslash 1_A, 3\backslash 3_A \end{matrix}\right|;$$

and this, by the theorem above cited,

$$= D_a^{2-1}.\, 2\backslash 2_a.(-1)^{2.(4+4)} = D_a.2\backslash 2_a;$$

$$\therefore \quad D_a = \frac{\left|\begin{matrix} 3\backslash 3_A, -3\backslash 1_A \\ -1\backslash 3_A, \ 1\backslash 1_A \end{matrix}\right|}{2\backslash 2_a};$$

which proves the method for a Block of 9 terms.

Next, let us take a Block of 16 terms, and represent it by

$$\left\{\begin{matrix} 1\backslash 1 \ldots\ldots 1\backslash 4 \\ \vdots \qquad\quad \vdots \\ 4\backslash 1 \ldots\ldots 4\backslash 4 \end{matrix}\right\}_a, \text{ and the adjugate Block by } \left\{\begin{matrix} 1\backslash 1 \ldots\ldots 1\backslash 4 \\ \vdots \qquad\quad \vdots \\ 4\backslash 1 \ldots\ldots 4\backslash 4 \end{matrix}\right\}_A.$$

If we 'condense' this, we get a Block of 9 terms; let us represent it by

$$\left\{\begin{matrix} 1\backslash 1 \ldots\ldots 1\backslash 3 \\ \vdots \qquad\quad \vdots \\ 3\backslash 1 \ldots\ldots 3\backslash 3 \end{matrix}\right\}_b,$$

so that $1\backslash 1_b = \begin{vmatrix} 1\backslash 1, 1\backslash 2 \\ 2\backslash 1, 2\backslash 2 \end{vmatrix}_a$, $1\backslash 2_b = \begin{vmatrix} 1\backslash 2, 1\backslash 3 \\ 2\backslash 2, 2\backslash 3 \end{vmatrix}_a$, &c.

If we 'condense' this Block again, we get a Block of 4 terms, each of which is, by the preceding paragraph, the Determinant of 9 terms of the Block of 16 terms;

that is, we get the Block $\left\{\begin{matrix} 4\backslash 4_A, -4\backslash 1_A \\ -4\backslash 1_A, \ 1\backslash 1_A \end{matrix}\right\}$;

and the Determinant of this will remain unchanged if we transpose the columns, and also the rows, and then multiply the first row and column by $-1$;

hence it $= \begin{vmatrix} 1\backslash 1_A, 1\backslash 4_A \\ 4\backslash 1_A, 4\backslash 4_A \end{vmatrix}$;

and this, by the theorem above cited,

$$= D_a^{2-1} . \begin{vmatrix} 2\backslash 2_a, 2\backslash 3_a \\ 3\backslash 2_a, 3\backslash 3_a \end{vmatrix} . (-1)^{2.(5+5)};$$

$$= D_a . 2\backslash 2_b;$$

$$\therefore \quad D_a = \frac{\begin{vmatrix} 4\backslash 4_A, -4\backslash 1_A \\ -1\backslash 4_A, \ 1\backslash 1_A \end{vmatrix}}{2\backslash 2_b};$$

which proves the method for a Block of 16 terms, and similar proofs might be given for larger Blocks.

# APPENDIX IV.

## Application of the method of 'Condensation' to the solution of simultaneous Linear Equations.

If we take a Block containing $n$ rows and $\overline{n+1}$ columns, and 'condense' it, we reduce it at last to 2 terms, the first of which is the Determinant of the first $n$ columns, the other of the last $n$ columns.

Hence, if we take the $n$ simultaneous Equations

$$1\backslash 1.x_1 + 1\backslash 2.x_2 + \ldots\ldots + 1\backslash n.x_n + 1\backslash \underline{n+1} = 0,$$

&c.

$$n\backslash 1.x_1 + n\backslash 2.x_2 + \ldots\ldots + n\backslash n.x_n + n\backslash \underline{n+1} = 0;$$

and if we 'condense' their $B$-Block, we reduce it to 2 terms, the first of which is $V$, the other $D_1$.

Now we know that $x_1 = (-)^n.\dfrac{D_1}{V}$; that is, $(-)^n V.x_1 = D_1$.

Hence the 2 terms obtained by the process of condensation may be converted into an Equation for $x_1$, by multiplying the first of them by $x_1$, affected with + or −, according as $n$ is even or odd. The latter part of the rule may be simply expressed thus:—"place the signs + and − alternately over the several columns, beginning with the last, and the sign which occurs over the column containing $x_1$ is the sign with which $x_1$ is to be affected."

When the value of $x_1$ has been thus found, it may be substituted in the first $\overline{n-1}$ Equations, and the same operation repeated on the new Block, which will now consist of $\overline{n-1}$ rows and $n$ columns. But in calculating the second series of Blocks, it will be found that most of the work has been already done; in fact, of the 2 Determinants required in the new Block, one has been already computed correctly, and the other so nearly so that it is only necessary to correct the *last* column in each of the derived Blocks.

In the example given opposite, after writing + and − alternately over the columns, beginning with the last, we first condense the whole Block, and thus obtain the 2 terms 36 and −72. Observing that the $x$-column has the sign − placed over it, we multiply the 36 by $-x$, and so form the Equation $-36x = -72$, which gives $x = 2$.

Hence the $x$-terms in the first four Equations become respectively 2, 2, 4, and 2; adding these values to the constant terms in the same Equations, we obtain a Block of which we need only write down the last two columns, viz. $\begin{vmatrix} 2 & 4 \\ -1 & -2 \\ -1 & -2 \\ 2 & 6 \end{vmatrix}$.

We then condense these into the column $\begin{vmatrix} 0 \\ 0 \\ 2 \end{vmatrix}$, and, supplying from the second Block of the first series the column $\begin{vmatrix} 3 \\ -1 \\ -5 \end{vmatrix}$, we obtain $\begin{vmatrix} 3 & 0 \\ -1 & 0 \\ -5 & 2 \end{vmatrix}$ as the last two columns of the *second* Block of the new series; and proceeding thus we ultimately obtain the two terms 12, 12. Observing that the $y$-column has the sign + placed over it, we multiply the first 12 by $+y$, and so form the Equation $12y = 12$, which gives $y = 1$. The values of $z$, $u$, and $v$ are similarly found.

It will be seen that when once the given Block has been successfully condensed, and the value of the first unknown obtained, there is no further danger of the operation being interrupted by the occurrence of ciphers.

$$\begin{array}{cccccc} - & + & - & + & - & + \end{array}$$

$$\begin{aligned} x + 2y + z - u + 2v + 2 &= 0 \\ x - y - 2z - u - v - 4 &= 0 \\ 2x + y - z - 2u - v - 6 &= 0 \\ x - 2y - z - u + 2v + 4 &= 0 \\ 2x - y + 2z + u - 3v - 8 &= 0 \end{aligned}$$

$$\begin{vmatrix} 1 & 2 & 1 & -1 & 2 & 2 \\ 1 & -1 & -2 & -1 & -1 & -4 \\ 2 & 1 & -1 & -2 & -1 & -6 \\ 1 & -2 & -1 & -1 & 2 & 4 \\ 2 & -1 & 2 & 1 & -3 & -8 \end{vmatrix} \quad \begin{vmatrix} 2 & 4 \\ -1 & -2 \\ -1 & -2 \\ 2 & 6 \end{vmatrix} \quad \begin{vmatrix} 2 & 6 \\ -1 & -3 \\ -1 & -1 \end{vmatrix} \quad \begin{vmatrix} 2 & 5 \\ -1 & -1 \end{vmatrix} \quad \begin{vmatrix} 2 & 4 \end{vmatrix}$$

$\therefore -2v = 4$ ∴ $v = -2$

$\begin{vmatrix} 3 & 3 \end{vmatrix}$ $\therefore 3u = 3$ …… ∴ $u = 1$

$$\begin{vmatrix} 3 & 0 \\ -1 & -2 \end{vmatrix} \quad \begin{vmatrix} 3 & 0 \\ -1 & 0 \\ -5 & -2 \end{vmatrix} \quad \begin{vmatrix} 6 & 6 \end{vmatrix}$$

$\therefore -6z = 6$ …………………… ∴ $z = -1$

$$\begin{vmatrix} -3 & -3 & -3 & 3 & -6 \\ 3 & 3 & 3 & -1 & 2 \\ -5 & -3 & -1 & -5 & 8 \\ 3 & -5 & 1 & 1 & -4 \end{vmatrix} \quad \begin{vmatrix} 6 & 0 \\ 8 & -2 \end{vmatrix}$$

$$\begin{vmatrix} 0 & 0 & 6 & 0 \\ 6 & -6 & 8 & -2 \\ -17 & 8 & -4 & 6 \end{vmatrix} \quad \begin{vmatrix} 12 & 12 \end{vmatrix}$$

$\therefore 12y = 12$ ………………………………… ∴ $y = 1$

$$\begin{vmatrix} 0 & 12 & 12 \\ 18 & 40 & -8 \end{vmatrix}$$

$\begin{vmatrix} 36 & -72 \end{vmatrix}$

$\because -36x = -72$ …………………………………………… ∴ $x = 2$

---

$$\begin{array}{cccc} - & + & - & + \end{array}$$

$$\begin{aligned} 5x + 2y - 3z + 3 &= 0 \\ 3x - y - 2z + 7 &= 0 \\ 2x + 3y + z - 12 &= 0 \end{aligned}$$

$$\begin{vmatrix} 5 & 2 & -3 & 3 \\ 3 & -1 & -2 & 7 \\ 2 & 3 & 1 & -12 \end{vmatrix} \quad \begin{vmatrix} -3 & 8 \\ -2 & 10 \end{vmatrix} \quad \begin{vmatrix} -3 & 12 \end{vmatrix}$$

∴ $3z = 12$ ……………… ∴ $z = 4$

$\begin{vmatrix} -7 & -14 \end{vmatrix}$

$\therefore -7y = -14$ …………………………………… ∴ $y = 2$

$$\begin{vmatrix} -11 & -7 & -15 \\ 11 & 5 & 17 \end{vmatrix}$$

$\begin{vmatrix} -22 & 22 \end{vmatrix}$

∴ $22x = 22$ ………………………………………………………… ∴ $x = 1$

# APPENDIX V.

SOLUTION OF THE PROBLEM "*Given an algebraic function of 2 or more terms; to construct a square Block, whose Elements are all mononomial, and whose Determinant vanishes simultaneously with the given function.*"

In what follows we shall use the word '*complemental*' in a somewhat extended sense: its definition may be given as follows: "*If, in a Block, any rows and any columns be selected: the Block formed of their common Elements, and the Block formed of the Elements common to the other rows and columns, are said to be* complemental *to each other.*" The definition previously given (CH. II. DEF. VII.) is evidently a particular case of this.

We have now to prove the following Theorem: "*If, in a square Block, any rows and any columns be selected; and if the Block formed of their common Elements be multiplied throughout by any quantity, and the complemental Block divided throughout by the same quantity: the Determinant of the new Block vanishes simultaneously with that of the first.*"

Call the degree of the first Block '$n$', its Determinant '$D$', the number of selected rows '$p$', of columns '$q$', and the quantity used for the processes of multiplication and division '$v$'.

First, let the selected $p$ rows be multiplied throughout by $v$; then the Determinant of the new Block, so formed, $= D.v^p$. (CH. II. AX. II.

Next, let the $\overline{n-q}$ columns, which were not selected, be divided throughout by $v$; then the Determinant of the new Block, so formed, $= D.v^{p+q-n}$; and therefore it vanishes simultaneously with $D$.

But, by the first process, the selected Block was multiplied throughout by $v$, as were also all the Elements common to the $p$ selected rows and the $\overline{n-q}$ columns which were not selected; and, by the second process, these latter Elements were again divided by $v$, as was also the Block complemental to the one selected.

Thus, by the two processes, the selected Block was multiplied throughout, and the complemental Block divided throughout, by $v$.

Hence this Theorem has been proved true.

Now in the square Block $\left\{\begin{matrix} 1\backslash 1 & 1\backslash 2 & 1\backslash 3 \\ 2\backslash 1 & 2\backslash 2 & 2\backslash 3 \\ 2\backslash 1 & 3\backslash 2 & 3\backslash 3 \end{matrix}\right\}$, let each of the Elements $2\backslash 2$, $3\backslash 3$, be divided, and its complemental Minor multiplied, by the Element itself. We thus obtain (neglecting exterior factors, which do not affect its evanescence) the Block $\left\{\begin{matrix} 1\backslash 1.2\backslash 2.3\backslash 3, & 1\backslash 2.3\backslash 3, & 1\backslash 3.2\backslash 2 \\ 2\backslash 1.3\backslash 3, & 3\backslash 3, & 2\backslash 3 \\ 2\backslash 2.3\backslash 1, & 3\backslash 2, & 2\backslash 2 \end{matrix}\right\}$. Then multiplying the 2nd row by $1\backslash 2$, the 3rd row by $1\backslash 3$, the 2nd columns by $2\backslash 1$, and the 3rd by $3\backslash 1$, we obtain $\left\{\begin{matrix} 1\backslash 1.2\backslash 2.3\backslash 3, & 1\backslash 2.2\backslash 1.3\backslash 3, & 1\backslash 3.2\backslash 2.3\backslash 1 \\ 1\backslash 2.2\backslash 1.3\backslash 3, & 1\backslash 2.2\backslash 1.3\backslash 3, & 1\backslash 2.2\backslash 3.3\backslash 1 \\ 1\backslash 3.2\backslash 2.3\backslash 1, & 1\backslash 3.2\backslash 1.3\backslash 2, & 1\backslash 3.2\backslash 2.3\backslash 1 \end{matrix}\right\}$, every Element of which is a Constituent of the first Block.

Now let it be given that $A$, $B$, $C$, $D$, $E$, $F$ are the 6 Constituents of a certain square Block of 9 terms: then we have the 6 Equations

$$\begin{aligned} A &= 1\backslash 1.2\backslash 2.3\backslash 3, & -D &= 1\backslash 1.2\backslash 3.3\backslash 2, \\ B &= 1\backslash 2.2\backslash 3.3\backslash 1, & -E &= 1\backslash 2.2\backslash 1.3\backslash 3, \\ C &= 1\backslash 3.2\backslash 1.3\backslash 2, & -F &= 1\backslash 3.2\backslash 2.3\backslash 1: \end{aligned}$$

then, substituting in the Block just found, we obtain $\left\{\begin{matrix} A, & -E, & -F \\ -E, & -E, & B \\ -F, & C, & -F \end{matrix}\right\}$, or multiplying the 1st row and 1st column by $-1$, $\left\{\begin{matrix} A, & E, & F \\ E, & -E, & B \\ F, & C, & -F \end{matrix}\right\}$.

The Determinant of this Block $= EF.\left\{A - \dfrac{ABC}{EF} + E + B + F + C\right\}$, and this, since $-\frac{ABC}{EF}=D$, becomes $EF.\{A+B+C+D+E+F\}$, and so is evanescent simultaneously with the unknown Block. And this condition, $ABC=-DEF$, is the only one which the 6 given quantities must fulfil, that the problem may be possible.

If the given Algebraical function contain 6 terms, we have only to apply this test, by grouping them into sets of 3, and if they satisfy the test, the Determinant can be written out at once: this may be done by multiplying together the 6 terms, taking the square root of their product, and finding, if possible, a set of three terms, whose product is equal to that square root, and whose sign is contrary to that of the product of the other 3 terms. Let us take as an example

$$5ab^2+3a^2-3abc-2ab+10a-4c;$$

here the continued product is $2.2.4.3.3.5.5a^6b^4c^2$, whose square root is $60a^3b^2c$, and this may be made up of the 3 terms $5ab^2$, $3a^2$, $-4c$; and as the sign of this product is $-$, and the sign of the product of the other 3 terms is $+$, the problem is possible. Hence, calling these 3 terms '$A, B, C$,' and the terms $10a$, $-2ab$, '$E, F$', we obtain the Block

$$\left\{\begin{matrix} 5ab^2, & 10a, & -2ab \\ 10a, & 10a, & 3a^2 \\ -2ab, & -4c, & 2ab \end{matrix}\right\},$$

which, if we divide rows by common factors, reduces to $\left\{\begin{matrix} 5b^2, & 10, & -2b \\ 10, & -10, & 3a \\ -ab, & -2c, & ab \end{matrix}\right\}$,

that is, dividing the 2nd column, to $\left\{\begin{matrix} 5b^2, & 5, & -2b \\ 10, & -5, & 3a \\ -ab, & -c, & ab \end{matrix}\right\}$, that is, multiplying the central term, and dividing its complemental Minor, by '$b$,' and also multiplying the last term in the 3rd row, and dividing its complemental Minor, by '5,' and changing the signs of the last term in the 1st row and of its complemental Minor $\left\{\begin{matrix} b, & 1, & 2 \\ -2, & b, & 3a \\ a, & c, & 5a \end{matrix}\right\}$.

If the given function contain 5 terms only, it will be necessary to break up one of them into 2 portions. In this case we ought to find 4 terms whose continuous product is such that the Algebraical portion of it is a square, and form them into 2 groups, each of which furnishes the square root of this product. We then break the 5th term into 2 portions, assigning one to each group, and in doing so we have only to attend to the numerical coefficients. As an example of this let us take

$$5a^4b-4a^3c+3a^2bc+2ac^2+11b^2;$$

here the continued product of the first 4 terms is $8.3.5.a^{10}b^2c^4$, and the square root of the Algebraical portion of this is $a^5bc^2$, and this is furnished by the product of $5a^4b$ and $2ac^2$: hence, arranging the terms in 2 groups, $5a^4b.2ac^2$ and $-4a^3c.3a^2bc$, we find by inspection that the last term must be broken into the 2 portions $6b^2$ and $5b^2$. Thus the two products, taken 3 and 3, become $5a^4b.2ac^2.6b^2$ and $-4a^3c.3a^2bc.5b^2$. Thus the required test is fulfilled and the Block may be written $\left\{\begin{matrix} 5a^4b, & -4a^3c, & 3a^2bc \\ -4a^3c, & 4a^3c, & 2ac^2 \\ 3a^2bc, & 6b^2, & -3a^2bc \end{matrix}\right\}$, which, dividing rows by common factors, becomes $\left\{\begin{matrix} 5a^2b, & -4ac, & 3bc \\ -2a^2, & 2a^2, & c \\ a^2c, & 2b, & -a^2c \end{matrix}\right\}$; and this again, dividing columns, becomes $\left\{\begin{matrix} 5b, & -2ac, & 3b \\ -2, & a^2 & 1 \\ c, & b, & -a^2 \end{matrix}\right\}$.

If the given function contain 4 terms only, we may proceed as in the case of 6, and append two equal terms with opposite signs: $+1$ and $-1$ are most convenient. For example, if the given function be

$$3a^2bc^2-4ab^4c-6ac^2+8b^3;$$

the continued product is $4.2.8.3.3.a^4b^8c^4$, whose square root is $24.a^2b^4c^2$, and this may be made up by the 2 terms $3a^2bc^2$ and $8b^3$. Hence the 6 terms may be taken to be $3a^2bc^2$, $8b^3$, 1, and $-4ab^4c$, $-6ac$, $-1$. Thus the test is fulfilled, and the Block may be written $\left\{\begin{matrix} 3a^2bc^2, & -6ac, & -1 \\ -6ac, & 6ac, & 8b^3 \\ -1, & 1, & 1 \end{matrix}\right\}$, that is, multiplying the 3rd term of the 3rd row, and dividing its complemental Minor, by $3ac$, and also dividing the 2nd row by 2, and changing the signs of the 1st row and 1st column, $\left\{\begin{matrix} abc, & 2, & 1 \\ 1, & 1, & 4b^3 \\ 1, & 1, & 3ac \end{matrix}\right\}$.

A Block of 16 terms may be constructed by a process similar to that employed for 9 terms.

Thus, in the Block $\left\{\begin{array}{cccc} 1\backslash 1 & 1\backslash 2 & 1\backslash 3 & 1\backslash 4 \\ 2\backslash 1 & 2\backslash 2 & 2\backslash 3 & 2\backslash 4 \\ 3\backslash 1 & 3\backslash 2 & 3\backslash 3 & 3\backslash 4 \\ 4\backslash 1 & 4\backslash 2 & 4\backslash 3 & 4\backslash 4 \end{array}\right\}$, let each of the Elements $2\backslash 2$, $3\backslash 3$, $4\backslash 4$, be divided, and its complemental Minor multiplied, by the Element itself. We thus obtain the Block

$$\left\{\begin{array}{cccc} 1\backslash 1.2\backslash 2.3\backslash 3.4\backslash 4, & 1\backslash 2.3\backslash 3.4\backslash 4, & 1\backslash 3.2\backslash 2.4\backslash 4, & 1\backslash 4.2\backslash 2.3\backslash 3 \\ 2\backslash 1.3\backslash 3.4\backslash 4, & 3\backslash 3.4\backslash 4, & 2\backslash 3.4\backslash 4, & 2\backslash 4.3\backslash 3 \\ 2\backslash 2.3\backslash 1.4\backslash 4, & 3\backslash 2.4\backslash 4, & 2\backslash 2.4\backslash 4, & 2\backslash 2.3\backslash 4 \\ 2\backslash 2.3\backslash 3.4\backslash 1, & 3\backslash 3.4\backslash 2, & 2\backslash 2.4\backslash 3, & 2\backslash 2.3\backslash 3 \end{array}\right\}.$$

Then, multiplying the 2nd row by $1\backslash 2$, the 3rd by $1\backslash 3$, the 4th by $1\backslash 4$, the 2nd column by $2\backslash 1$, the 3rd by $3\backslash 1$, and the 4th by $4\backslash 1$, we obtain the Block

$$\left\{\begin{array}{cccc} 1\backslash 1.2\backslash 2.3\backslash 3.4\backslash 4, & 1\backslash 2.2\backslash 1.3\backslash 3.4\backslash 4, & 1\backslash 3.2\backslash 2.3\backslash 1.4\backslash 4, & 1\backslash 4.2\backslash 2.3\backslash 3.4\backslash 1 \\ 1\backslash 2.2\backslash 1.3\backslash 3.4\backslash 4, & 1\backslash 2.2\backslash 1.3\backslash 3.4\backslash 4, & 1\backslash 2.2\backslash 3.3\backslash 1.4\backslash 4, & 1\backslash 2.2\backslash 4.3\backslash 3.4\backslash 1 \\ 1\backslash 3.2\backslash 2.3\backslash 1.4\backslash 4, & 1\backslash 3.2\backslash 1.3\backslash 2.4\backslash 4, & 1\backslash 3.2\backslash 2.3\backslash 1.4\backslash 4, & 1\backslash 3.2\backslash 2.3\backslash 4.4\backslash 1 \\ 1\backslash 4.2\backslash 2.3\backslash 3.4\backslash 1, & 1\backslash 4.2\backslash 1.3\backslash 3.4\backslash 2, & 1\backslash 4.2\backslash 2.3\backslash 1.4\backslash 3, & 1\backslash 4.2\backslash 2.3\backslash 3.4\backslash 1 \end{array}\right\},$$

every Element of which is a Constituent of the 1st Block.

Now let it be given that $A$, $B$, $C$, &c. are the 24 Constituents of a certain square Block of 16 terms: then we have the 24 Equations

$$\left.\begin{array}{l} A = 1\backslash 1.2\backslash 2.3\backslash 3.4\backslash 4 \\ B = 1\backslash 2.2\backslash 1.3\backslash 4.4\backslash 3 \\ C = 1\backslash 3.2\backslash 4.3\backslash 1.4\backslash 2 \\ D = 1\backslash 4.2\backslash 3.3\backslash 2.4\backslash 1 \end{array}\right\}, \quad \left.\begin{array}{l} E = 1\backslash 1.2\backslash 3.3\backslash 4.4\backslash 2 \\ F = 1\backslash 2.2\backslash 4.3\backslash 3.4\backslash 1 \\ G = 1\backslash 3.2\backslash 1.3\backslash 2.4\backslash 4 \\ H = 1\backslash 4.2\backslash 2.3\backslash 1.4\backslash 3 \end{array}\right\}, \quad \left.\begin{array}{l} J = 1\backslash 1.2\backslash 4.3\backslash 2.4\backslash 3 \\ K = 1\backslash 2.2\backslash 3.3\backslash 1.4\backslash 4 \\ L = 1\backslash 3.2\backslash 2.3\backslash 4.4\backslash 1 \\ M = 1\backslash 4.2\backslash 1.3\backslash 3.4\backslash 2 \end{array}\right\},$$

$$\left.\begin{array}{l} N = 1\backslash 1.2\backslash 2.3\backslash 4.4\backslash 3 \\ P = 1\backslash 2.2\backslash 1.3\backslash 3.4\backslash 4 \\ -Q = 1\backslash 3.2\backslash 4.3\backslash 2.4\backslash 1 \\ -R = 1\backslash 4.2\backslash 3.3\backslash 1.4\backslash 2 \end{array}\right\}, \quad \left.\begin{array}{l} -S = 1\backslash 1.2\backslash 4.3\backslash 3.4\backslash 2 \\ -T = 1\backslash 2.2\backslash 3.3\backslash 4.4\backslash 1 \\ -U = 1\backslash 3.2\backslash 2.3\backslash 1.4\backslash 4 \\ -V = 1\backslash 4.2\backslash 1.3\backslash 2.4\backslash 3 \end{array}\right\}, \quad \left.\begin{array}{l} -W = 1\backslash 1.2\backslash 3.3\backslash 2.4\backslash 4 \\ -X = 1\backslash 2.2\backslash 4.3\backslash 1.4\backslash 3 \\ -Y = 1\backslash 3.2\backslash 1.3\backslash 4.4\backslash 2 \\ -Z = 1\backslash 4.2\backslash 2.3\backslash 3.4\backslash 1 \end{array}\right\},$$

then, substituting in the Block just found, we obtain $\left\{\begin{matrix} A, -P, -U, -Z \\ -P, -P, K, F \\ -U, G, -U, L \\ -Z, M, H, Z \end{matrix}\right\}$,

which, if we change the signs of the 1st row and 1st column, becomes

$$\left\{\begin{matrix} A, & P, & U, & Z \\ P, & -P, & K, & F \\ U, & G, & -U, & L \\ Z, & M, & H, & -Z \end{matrix}\right\}.$$

This contains 10 only of the given 24 quantities, so that, in order to prove that its Determinant contains $(A+B+\&c.+Z)$ as a factor, we must have 14 independent relations among the given quantities.

The 10 quantities which enter into the above Block are

$$A, F, G, H, K, L, M, P, U, Z;$$

and the following Equations give the remaining 14 quantities in terms of these:—

$$\left.\begin{matrix} B = -\frac{HLP}{UZ} \\ C = -\frac{FMU}{PZ} \\ D = -\frac{GKZ}{PU} \end{matrix}\right\}, \left.\begin{matrix} E = -\frac{AKLM}{PUZ} \\ J = -\frac{AFGH}{PUZ} \end{matrix}\right\}, \left.\begin{matrix} N = -\frac{AHL}{UZ} \\ Q = -\frac{FG}{P} \\ R = -\frac{KM}{P} \end{matrix}\right\}, \left.\begin{matrix} S = -\frac{AFM}{PZ} \\ T = -\frac{KL}{U} \\ V = -\frac{GH}{U} \end{matrix}\right\}, \left.\begin{matrix} W = -\frac{AGK}{PU} \\ X = -\frac{FH}{Z} \\ Y = -\frac{LM}{Z} \end{matrix}\right\}$$

# TABULAR VIEW OF ANALYSIS OF EQUATIONS. (CHAP. III.)

## I. *n Equations not all homogeneous.*

| | DATA. | | | | QUÆSITA. | | |
|---|---|---|---|---|---|---|---|
| | *No. of Variables* | *V-Block* | *B-Block* | *Other data* | *Consistency of Equations* | *Interdependency of Equations* | *Variables to which arbitrary values may be assigned.* |
| PROP. I. | $n$ | $V \neq 0$ | | | consistent | all independent | none |
| PROP. II. | $n+r$ | $\lvert V \rvert \neq 0$ | | | — | — | if any non-evanescent principal Minor of the $V$-Block be selected, the $r$ Variables, whose coefficients are not contained in it, may have arbitrary values assigned to them |
| PROP. III. | $n-1$ | | $B \neq 0$ | | inconsistent | | |
| COR. | $n-r$ | | $\lvert B \rvert \neq 0$ | | — | | |
| PROP. IV. | $n$ | $V = 0$ | $\lvert B \rvert \neq 0$ | | — | | |
| PROP. V. | $n+r$ | $\lVert V \rVert = 0$ | — | | — | | |

| | | | | | | | |
|---|---|---|---|---|---|---|---|
| Prop. VIII. | $n-1$ | | $B = 0$ | there are among them $\overline{n-1}$ Equations which have their $V$-Block not evanescent | consistent | the remaining Equation is dependent on these $\overline{n-1}$ Equations | |
| Cor. | $n-r$ | | — | there are among them $\overline{n-r}$ Equations which have their $V$-Block not evanescent, and, when these are taken along with each of the remaining Equations successively, each set, so formed, has its $B$-Block evanescent | — | the remaining $r$ Equations are dependent on these $\overline{n-r}$ Equations. | — |
| Prop. IX. | $n$ | | $\|B\| = 0$ | there are among them $\overline{n-1}$ Equations which have their $V$-Block not evanescent | consistent | the remaining Equation is dependent on these $\overline{n-1}$ Equations | if any non-evanescent principal Minor of the $V$-Block of these $\overline{n-1}$ Equations be selected, the Variables, whose coefficients are not contained in it, may have an arbitrary value assigned to it |
| Cor. | $n+r$ | | — | — | — | do. for $\overline{r+1}$ Equations | do. for $\overline{r+1}$ Variables |
| Prop. X. | $n$ | | | there are among them $\overline{n-k}$ Equations which have their $V$-Block not evanescent, and, when these are taken along with each of the remaining Equations successively, each set, so formed, has its $B$-Block evanescent | consistent | the remaining $k$ Equations are dependent on these $\overline{n-k}$ Equations | if any non-evanescent principal Minor of these $\overline{n-k}$ Equations be selected, the $k$-Variables, whose coefficients are not contained in it, may have arbitrary values assigned to them |
| Cor. | $n+r$ | | | — | — | do. for $\overline{k+r}$ Equations | do. for $\overline{k+r}$ Variables. |

# TABULAR VIEW OF ANALYSIS OF EQUATIONS. (Chap. III.)

## II. *n Equations all homogeneous.*

| | Data. | | | Quæsita. | | |
|---|---|---|---|---|---|---|
| | *No. of Variables* | *V-Block* | *Other data* | *Possibility of assigning to the Variables a set of values which are not all zero* | *Interdependence of Equations* | *Variables to which arbitrary values may be assigned* |
| Prop. I. | $n$ | $V \neq 0$ | | impossible | | |
| | $n-r$ | $\|V\| \neq 0$ | | — | | |
| Prop. II. | $n+1$ | $\|V\| \neq 0$ | | possible | all independent | if any non-evanescent principal Minor of the $V$-Block be selected, the Variable, whose coefficients are not contained in it, may have an arbitrary value assigned to it |
| do. | $n+r$ | — | | — | — | do. for $\overline{r+1}$ Variables |

| | | | | | | |
|---|---|---|---|---|---|---|
| Prop. IX. | $n$ | $V=0$ | there are among them $\overline{n-1}$ Equations, which have their $V$-Block not evanescent | possible | the remaining Equation is dependent on these $\overline{n-1}$ Equations | if any non-evanescent principal Minor of the $V$-Block of these $\overline{n-1}$ Equations be selected, the Variable, whose coefficients are not contained in it, may have an arbitrary value assigned to it |
| Cor. | $n+r$ | — | — | — | — | do. for $\overline{r+1}$ Variables |
| Prop. X. | $n$ | $V=0$ | there are among them $\overline{n-k}$ Equations, which have their $V$-Block not evanescent, and, when these are taken along with each of the remaining Equations successively, each set, so formed, has its $V$-Block evanescent. | possible | the remaining $k$ Equations are dependent on these $\overline{n-k}$ Equations | if any non-evanescent principal Minor of these $\overline{n-k}$ Equations be selected, the $k$-Variables, whose coefficients are not contained in it, may have arbitrary values assigned to them |
| Cor. | $n+r$ | — | — | — | — | do. for $\overline{k+r}$ Variables |

# FORMULÆ.

## PART I.—*ALGEBRAICAL.*

| *Data.* | *Quæsita.* | |
|---|---|---|
| $n$ Equations, not all homogeneous, containing Variables | a test of their being consistent | either there is one of them such that, when it is taken along with each of the remaining Equations successively, each pair of Equations, so formed, has its $B$-Block evanescent; or there are $m$ of them, where $m$ is one of the numbers $2 \ldots\ldots n$, which contain at least $m$ Variables, and have their $V$-Block not evanescent, and are such that, when they are taken along with each of the remaining Equations successively, each set of Equations, so formed, has its $B$-Block evanescent. |
| 2 Equations containing Variables | a test of their being identical | $\|B\| = 0$. |
| $n$ homogeneous Equations, containing not more than $n$ Variables | a test of their being, for the Variables, a set of values, which are not all zero. | $\|V\| = 0$. |

| | | |
|---|---|---|
| An oblong Block, having one of its secondary Minors not evanescent | a test of its being evanescent | of the principal Minors, each one, which contains that secondary Minor, is evanescent. |
| A Block | a test of its being evanescent | either every Element of it is zero, or there are 2 or more of its longitudinals, which form a Block, having one of its secondary Minors not evanescent, and such that, of its principal Minors, each one, which contains that secondary Minor, is evanescent. |
| A Block | a test for the evanescence of every oblong Block, formed from it by selecting $h$ of its laterals, or longitudinals | the test for *laterals* is either that every Element of the Block is zero, or that it has a non-evanescent Minor of the $k^{th}$ degree, where $k$ is less than $h$, such that, of the oblong Blocks formed from it by selecting $\overline{k+1}$ of its laterals, each one, which contains that non-evanescent Minor, is evanescent : and the test for *longitudinals* is similar to this. |
| A function of 6 terms, fulfilling the condition $ABC = -DEF$ | a square Block of 9 terms, whose Elements are mononomial, and whose Determinant vanishes simultaneously with the given function | $\left\{\begin{matrix} A, & E, & F \\ E, & -E, & B \\ F, & C, & -F \end{matrix}\right\}.$ |

T 2

# FORMULÆ.

## PART II.—*GEOMETRICAL.*

## SECTION I.

### *Plane Geometry.*

| | *Cartesian.* | *Trilinear.* |
|---|---|---|
| Let given Points be represented by | $(x_1, y_1)$, $(x_2, y_2)$, &c. | $(\alpha_1, \beta_1, \gamma_1)$, $(\alpha_2, \beta_2, \gamma_2)$, &c. |
| Let given Lines be represented by | $A_1x + B_1y + C_1 = 0$, &c. | $A_1\alpha + B_1\beta + C_1\gamma + D_1 = 0$, &c. the systematic Equation being $a\alpha + b\beta + c\gamma - 2M = 0$. |

| *Data.* | *Quæsita.* | | |
|---|---|---|---|
| 2 Lines | test for their having the same direction | $\begin{vmatrix} A_1, B_1 \\ A_2, B_2 \end{vmatrix} = 0.$ | $\begin{vmatrix} A_2, B_1, C_1 \\ A_2, B_2, C_2 \\ a, b, c \end{vmatrix} = 0.$ |
| 3 Lines | test for their intersecting in one Point, at a finite or infinite distance | $\begin{vmatrix} A_1, B_1, C_1 \\ A_2, B_2, C_2 \\ A_3, B_3, C_3 \end{vmatrix} = 0.$ | $\begin{vmatrix} A_1, B_1, C_1, D_1 \\ A_2, B_2, C_2, D_2 \\ A_3, B_3, C_3, D_3 \\ a, b, c, -2M \end{vmatrix} = 0.$ |
| | further test for their so intersecting at a finite distance | that either there are, among the Equations to the 3 Lines, 2 which have their *V-Block* not evanescent, or else every 2 of them have their B-Block evanescent. | |
| 3 Points | test for their lying on one Line | $\begin{vmatrix} x_1, y_1, 1 \\ x_2, y_2, 1 \\ x_3, y_3, 1 \end{vmatrix} = 0.$ | $\begin{vmatrix} \alpha_1, \beta_1, \gamma_1 \\ \alpha_2, \beta_2, \gamma_2 \\ \alpha_3, \beta_3, \gamma_3 \end{vmatrix} = 0.$ |
| 2 Points | Equation to Line through them | $\begin{vmatrix} x, y, 1 \\ x_1, y_1, 1 \\ x_2, y_2, 1 \end{vmatrix} = 0.$ | $\begin{vmatrix} \alpha, \beta, \gamma \\ \alpha_1, \beta_1, \gamma_1 \\ \alpha_2, \beta_2, \gamma_2 \end{vmatrix} = 0.$ |

# SECTION II.

## *Solid Geometry.*

| Data. | Quæsita. | Cartesian. | Quadriplanar. |
|---|---|---|---|
| Let given Points be represented by | | $(x_1, y_1, z_1), (x_2, y_2, z_2)$, &c. | $(a_1, \beta_1, \gamma_1, \delta_1), (a_2, \beta_2, \gamma_2, \delta_2)$, &c. |
| Let given Planes be represented by | | $A_1x + B_1y + C_1z + D_1 = 0$, &c. | $A_1a + B_1\beta + C_1\gamma + D_1\delta + E_1 = 0$, &c. the systematic Equation being $aa + b\beta + c\gamma + d\delta - 3M = 0$. |
| 2 Planes | test for their having the same direction | $\left\| \begin{matrix} A_1, & B_1, & C_1 \\ A_2, & B_2, & C_2 \end{matrix} \right\| = 0.$ | $\left\| \begin{matrix} A_1, & B_1, & C_1, & D_1 \\ A_2, & B_2, & C_2, & D_2 \\ a, & b, & c, & d \end{matrix} \right\| = 0.$ |
| 3 Planes | test for their intersecting in one Line, at a finite or infinite distance | $\left\| \begin{matrix} A_1, & B_1, & C_1, & D_1 \\ A_2, & B_2, & C_2, & D_2 \\ A_3, & B_3, & C_3, & D_3 \end{matrix} \right\| = 0.$ | $\left\| \begin{matrix} A_1, & B_1, & C_1, & D_1, & E_1 \\ A_2, & B_2, & C_2, & D_2, & E_2 \\ A_3, & B_3, & C_3, & D_3, & E_3 \\ a, & b, & c, & d, & -3M \end{matrix} \right\| = 0.$ |
| | further test for their so intersecting at a finite distance | either there are, among the Equations to the 3 Planes, 2 which have their *V*-Block not evanescent, or else every 2 of them have their *B*-Block evanescent. | |

| | | | |
|---|---|---|---|
| 4 Planes | test for their intersecting in one Point, at a finite or infinite distance | $\begin{vmatrix} A_1, & B_1, & C_1, & D_1 \\ \vdots & \vdots & \vdots & \vdots \\ A_4, & B_4, & C_4, & D_4 \end{vmatrix} = 0.$ | $\begin{vmatrix} A_1, & B_1, & C_1, & D_1, & E_1 \\ \vdots & \vdots & \vdots & \vdots & \vdots \\ A_4, & B_4, & C_4, & D_4, & E_4 \\ a, & b, & c, & d, & -3M \end{vmatrix} = 0.$ |
| | further test for their so intersecting at a finite distance | either there are, among the Equations to the 4 Planes, 3 which have their $V$-Block not evanescent, or else every 3 have their $B$-Block evanescent and there are 2 which have their $V$-Block not evanescent, or else every 2 have their $B$-Block evanescent. | |
| 3 Points | test for their lying on one Line | $\begin{Vmatrix} x_1, & y_1, & z_1, & 1 \\ x_2, & y_2, & z_2, & 1 \\ x_3, & y_3, & z_3, & 1 \end{Vmatrix} = 0.$ | $\begin{Vmatrix} \alpha_1, & \beta_1, & \gamma_1, & \delta_1 \\ \alpha_2, & \beta_2, & \gamma_2, & \delta_2 \\ \alpha_3, & \beta_3, & \gamma_3, & \delta_3 \end{Vmatrix} = 0.$ |
| 4 Points | test for their lying on one Plane | $\begin{vmatrix} x_1, & y_1, & z_1, & 1 \\ \vdots & \vdots & \vdots & \vdots \\ x_4, & y_4, & z_4, & 1 \end{vmatrix} = 0.$ | $\begin{vmatrix} \alpha_1, & \beta_1, & \gamma_1, & \delta_4 \\ \vdots & \vdots & \vdots & \vdots \\ \alpha_4, & \beta_4, & \gamma_4, & \delta_4 \end{vmatrix} = 0.$ |
| 2 Points | Equation to Line through them | $\begin{Vmatrix} x, & y, & z, & 1 \\ x_1, & y_1, & z_1, & 1 \\ x_2, & y_2, & z_2, & 1 \end{Vmatrix} = 0.$ | $\begin{Vmatrix} \alpha, & \beta, & \gamma, & \delta \\ \alpha_1, & \beta_1, & \gamma_1, & \delta_1 \\ \alpha_2, & \beta_2, & \gamma_2, & \delta_2 \end{Vmatrix} = 0.$ |
| 3 Points | Equation to Plane through them | $\begin{vmatrix} x, & y, & z, & 1 \\ x_1, & y_1, & z_1, & 1 \\ x_2, & y_2, & z_2, & 1 \\ x_3, & y_3, & z_3, & 1 \end{vmatrix} = 0.$ | $\begin{vmatrix} \alpha, & \beta, & \gamma, & \delta \\ \alpha_1, & \beta_1, & \gamma_1, & \delta_1 \\ \alpha_2, & \beta_2, & \gamma_2, & \delta_2 \\ \alpha_3, & \beta_3, & \gamma_2, & \delta_3 \end{vmatrix} = 0.$ |

www.ingramcontent.com/pod-product-compliance
Lightning Source LLC
Chambersburg PA
CBHW021931190326
41519CB00009B/983
* 9 7 8 1 4 7 3 3 1 5 9 5 2 *